Insurgent Terrorism

CAUSES AND CONSEQUENCES OF TERRORISM SERIES

Series Editors:
Gary LaFree
Gary A. Ackerman
Anthony Lemieux

BOOKS IN THE SERIES:

From Freedom Fighters to Jihadists: Human Resources of Non State Armed Groups
Vera Mironova

ISIS Propaganda: A Full-Spectrum Extremist Message
Edited by Stephane J. Baele, Katharine A. Boyd, and Travis G. Coan

Extremist Islam: Recognition and Response in Southeast Asia
Kumar Ramakrishna

Insurgent Terrorism: Intergroup Relationships and the Killing of Civilians
Victor Asal, Brian J. Phillips, and R. Karl Rethemeyer

Published in partnership with The National Consortium for the Study of Terrorism and Responses to Terrorism and the University of Maryland

Insurgent Terrorism

*Intergroup Relationships and
the Killing of Civilians*

VICTOR ASAL, BRIAN J. PHILLIPS, AND
R. KARL RETHEMEYER

OXFORD

UNIVERSITY PRESS

OXFORD
UNIVERSITY PRESS

Oxford University Press is a department of the University of Oxford. It furthers
the University's objective of excellence in research, scholarship, and education
by publishing worldwide. Oxford is a registered trade mark of Oxford University
Press in the UK and certain other countries.

Published in the United States of America by Oxford University Press
198 Madison Avenue, New York, NY 10016, United States of America.

© University of Maryland National Consortium for the Study of
Terrorism and Responses to Terrorism (START) 2022

Library of Congress Cataloging-in-Publication Data
Names: Asal, Victor, author. | Phillips, Brian J., author. | Rethemeyer, R. Karl, author.
Title: Insurgent terrorism : intergroup relationships and the killing of
civilians / Victor Asal, Brian J. Phillips, R. Karl Rethemeyer.
Description: New York, NY : Oxford University Press, [2022] |
Series: Causes and consequences of terrorism | Includes bibliographical references and index.
Identifiers: LCCN 2021052801 (print) | LCCN 2021052802 (ebook) |
ISBN 9780197607015 (hardback) | ISBN 9780197607060 (paperback) |
ISBN 9780197607039 (epub) | ISBN 9780197607046 (digital-online)
Subjects: LCSH: Terrorism. | Insurgency. | Violence.
Classification: LCC HV6431 .A8226 2022 (print) | LCC HV6431 (ebook) |
DDC 363.325—dc23/eng/20211202
LC record available at https://lccn.loc.gov/2021052801
LC ebook record available at https://lccn.loc.gov/2021052802

DOI: 10.1093/oso/9780197607015.001.0001

Contents

Acknowledgments   vii

SECTION I. INTRODUCTION, THEORY, AND INITIAL TESTING

1. Introduction   3

2. The Embeddedness Theory of Civilian Targeting
   by Insurgent Organizations   21

3. Describing the Big, Allied, and Dangerous II Insurgency
   Data and Other Data Sources   45

4. Testing Primary Hypotheses   65

SECTION II. EMPIRICAL EXTENSIONS: TYPES OF CIVILIAN TARGETING

5. Why Do Some Insurgent Groups Mostly Attack
   the General Public?   85

6. Why Do Some Insurgent Groups Attack Schools?   107

7. Why Do Some Insurgent Groups Attack Journalists?   131

SECTION III. FURTHER ANALYSIS OF INTERGROUP RELATIONSHIPS

8. Longitudinal Modeling of Insurgent Alliances   155

9. Understanding Insurgent Rivalry   189

10. Conclusion   215

Index   229

Acknowledgments

This book is about insurgents embedded in networks of other actors. We, the authors, are also enmeshed in networks of colleagues, family members, institutions, funders, and other supporters without whom the book would not have been possible. It is the culmination of years of work, rooted in the original Big, Allied, and Dangerous (BAAD) project led by Victor and Karl, which focused on terrorist organizations and began in the early 2000s. The next step began when Victor and Karl started gathering data for the BAAD Insurgency project. This was funded primarily by the National Consortium for the Study of Terrorism and Responses to Terrorism (START) at the University of Maryland.[1] We are grateful to START for their support and the network of colleagues and collaborators who informed our thinking over more than 15 years.

In 2010, Victor agreed to be the external member of Brian's doctoral dissertation committee. (Burcu Savun introduced them, and they are grateful for that and for her excellent mentorship of Brian.) Victor, Karl, and Brian shared an interest in studying terrorism, insurgency, and networks, so we eventually started to work together on several research projects. This book is a product of a powerful collaboration in which all three of us have profited from the expertise and drive each of us brings to the enterprise.

We acknowledge the painstaking work by our coders to collect, perfect, and manage the BAAD data through their work at the University at Albany's Project on Violent Conflict. We are particularly indebted to Suzanne Weedon Levy, Corina Simonelli, Marcus Schultzke, James Levy, Ahkeel Owens, and Lois Pownall, who helped run the program and manage the finances. We thank all the student coders who collected data. The names of undergraduate

[1] This material is based on work supported by the following grants: US Army Research Laboratory (W911NF19F0014) under subcontract from the University of Maryland, US Department of Homeland Security (2012-ST-061-CS0001-05 and 2008-ST-061-ST0004) under subcontract from the University of Maryland, and Defense Threat Reduction Agency (HDTRA1-10-1-0017) under subcontract from the University of Arizona/Arizona Board of Regents. The views and conclusions contained in this book are those of the authors and should not be interpreted as necessarily representing the official policies, either expressed or implied, of the funding agencies or primary contractors.

and graduate coders appear later. Without out coders, we would never have been able to do this research. We hope they also profited from working on this project.

Although the embeddedness theory and most of the work in the book are being presented for the first time, some ideas in Section I originally appeared in an article co-authored with Corina Simonelli and Joseph K. Young [Asal, Victor, Brian J. Phillips, R. Karl Rethemeyer, Corina Simonelli, and Joseph K. Young. "Carrots, Sticks, and Insurgent Targeting of Civilians." *Journal of Conflict Resolution* 63, no. 7 (2019): 1710–35]. That paper served as a springboard for thinking about additional reasons behind why insurgent groups target civilians.

We appreciate the careful and hard-working team at Oxford: Abby Gross, Nadina Persaud, and Katharine Pratt. We also thank the reviewers of the manuscript, who provided valuable feedback that helped shape the final product. Special thanks also go to the START series editors: Gary Ackerman, Gary LaFree, and Anthony Lemieux.

A number of colleagues were generous enough to write blurbs for the book: Navin Bapat, Kathleen Gallagher Cunningham, Ursula Daxecker, Hanne Fjelde, Jim Piazza, and Todd Sandler. We thank them for their time and kind words.

Finally, we thank our families. Brian thanks his wife, Carolina, for her immeasurable support, including thoughtful discussions about the contents of this book. He also thanks Mia and Lucas for being their amazing and lovable selves. Karl thanks his wife, Jodi, and son, Ben, who all too often were told dinner would have to wait while the estimations finished running. Victor thanks his family as well.

The book is dedicated to victims of terrorism everywhere, with the hope that our contribution will help in at least some small way to direct political grievances toward peaceful resolutions that enhance safety, security, justice, and freedom for all.

Victor Asal, Albany, NY, USA
Brian J. Phillips, Wivenhoe, England
R. Karl Rethemeyer, Amherst, MA, USA

Big, Allied, and Dangerous Coders 2009–2021

Undergraduate Students

Kady Arthur, Lauren Bailey, Susan Bitter, Emily Blakeslee, James Brennan, Cherie Brown, Catherine Callahan, Alejandro Castro-Reina, Kevin Chambers, Natalie Cohen, Sean Correia, Charles Coryer, Stephanie Dandaraw, Paige Donegan, Kimberly Dorner, Danielle Duguid, Trevor Eck, Jacob Egloff, Emily Finnegan, Vincent Giannone, Daniel Gustafson, Kayla Hamburg, Chamberlain Harris, Meghan Hart, Jona Hoxha, Christian Johnson, Cassandra Jones, Brian Junquera, Morgan Knudtsen, Kristin Kopach, Michael Kriesberg, Serae LaFache-Brazier, Robin Lieb, Julia Marshall, Joel Murray, Jane Ni, Brianna Nugent, Ahkeel Owens, Joshua Rosenstein, Casey Schur, Corina Simonelli, Ariel Spielman, Brandon Suter, Francesca Tse, Sean Whipple, Kristina Wieneke, and Briana Young.

Graduate Students

Amira Amari, Emily Blakeslee, Jeffrey Blauvelt, Jr., Katelyn Bracht, Shannon Briggs, Nolan Fahrenkopf, Adil Fala, Vincent Giannone, Amira Jadoon, Nakissa Jahanbani, Jayson Kratoville, Hyun-Hee Park, Huidong Peng, Marcus Schulzke, Katherine Slye, Sean Stephens, Yi-hao Su, Andrew Vitek, and Suzanne Weedon.

SECTION I
INTRODUCTION, THEORY, AND INITIAL TESTING

1

Introduction

1.1. Introduction

Imagine getting on a bus to travel from one major city to another. It had been a long week and all you want to do is get home and take a nap during the bus ride. Imagine falling asleep and enjoying the rest on the bus. Now imagine that as the bus is driving up a mountain, you wake to someone screaming incoherently and you can feel the bus swerve to the right and through a road barrier and over the side of the mountain. Some of the people you are with on the bus fly out the window as it crashes down the mountain into a ravine while others fly around the bus slamming into each other, into metal, and into shattering glass. As the bus slams down, you can feel parts of your body break, and you see other people die in front of you. You then lose consciousness. When you wake, you are lying outside the bus with glass and screaming people around you. The roof of the bus is now on the ground. You are in severe pain as you look around and see dead, dying, and broken people all around you and dozens of individuals streaming down the valley to help you and the other injured victims.

This is not an imagined story but something that happened on a Thursday in the summer of 1989 on the 405 bus that drove regularly between Tel Aviv and Jerusalem. A Palestinian, Abd al-Hadi Ghanim, grabbed the wheel from the driver and steered the bus off the side of a cliff. Sixteen civilians were killed in the attack—some from being thrown out of the bus and some from being burned to death when part of the bus caught fire. More than 25 other passengers were injured. Although Ghanim did not die in the crash, his assault was viewed as the first suicide attack by Palestinians in Israel and the occupied territories (Zieve 2012). Ghanim was acting on behalf of the Palestinian Islamic Jihad, a militant organization that only in the 2000s started killing enough soldiers to enter into the Uppsala Conflict Data Program (UCDP) battle deaths data (Gleditsch et al. 2002; Pettersson, Högbladh, and Öberg 2019). In the 1980s and 1990s, however, the group was primarily focused on attacking civilians (BBC 2003).

Insurgent Terrorism. Victor Asal, Brian J. Phillips, and R. Karl Rethemeyer, Oxford University Press. © University of Maryland National Consortium for the Study of Terrorism and Responses to Terrorism (START) 2022. DOI: 10.1093/oso/9780197607015.003.0001

Weak organizations—those unable to kill soldiers on a regular basis—are perhaps what most observers think of as groups that use terrorism. But groups such as the Palestinian Islamic Jihad in the late 1980s are not the only type of organization that has focused on killing civilians. In the last weeks of 2011 and the first weeks of 2012, the organization Boko Haram killed more than 37 people—all civilians (Mshelizza 2012). During a 2-day period, Boko Haram killed at least 6 worshipers in the Deeper Life Church in Gombe State in Nigeria and an 80-year-old and his son in the state of Borno (Amaize and Yusuf 2012). The organization also killed people in a church in Adamawa State. A spokesperson for the group claimed that it was responsible for an explosion on Christmas Day near a church on December 25, 2011, and then threatened more violence (Amaize and Yusuf 2012). The horror of these attacks is starkly clear in what happened to the Adurkwa family in northern Nigeria (Freeman 2012):

> On Wednesday evening, three days after Boko Haram ordered all Christians to leave Muslim-dominated northern Nigeria, Ousman Adurkwa, a 65-year-old local trader, answered the door of his home near the church to what he thought was an after-hours customer. Instead it was two masked gunmen.
>
> "They shot my father dead, and then came for the rest of the family," Adurkwa's son Hyeladi, 25, told *The Sunday Telegraph* the following day. "One chased my brother Moussa and killed him, and the other shot at me, but my mother took the bullet in the stomach instead."
>
> Hyeladi spoke as weeping parishioners gathered for an impromptu memorial service in the Adurkwa family compound, where the parlour carpet was still stained with blood from the gunshot wound suffered by Mrs. Aduwurka, 50, who now lies in hospital.

In response to these attacks, and a Boko Haram warning that Christians had 3 days to leave northern Nigeria, hundreds of Christians fled to the south and reprisals against Muslims were expected (Mshelizza 2012). This killing of civilians would make Boko Harm easily fit many definitions of terrorist organizations (Asal et al. 2012). However, during this period, Boko Haram was not just an organization that killed civilians. Since 2009, Boko Haram has waged war against security forces of Nigeria and other countries, resulting in hundreds or thousands of "battle deaths" almost every year, according to UCDP. The group is widely described as an insurgent organization, actively waging insurgency (Council on Foreign Relations 2018; Osumah 2013).

Another organization well known for civilian targeting that has also been involved in a great deal of combat against government soldiers is the so-called Islamic State or ISIS. For example, on January 8, 2007, Iraqi soldiers organized an attack on ISIS (then called al-Qaeda in Iraq) and other Iraqi insurgent organizations to take back control of Haifa Street in Baghdad. ISIS fought off the attack, and the Iraqi soldiers called in U.S. troops (Kagan 2008; Roggio 2007). When U.S. forces launched their attack, ISIS militants launched a fierce and sustained counterattack (Kagan 2008):

> The 1-23rd Infantry Battalion moved out of its assembly area at 3:30 a.m. on Tuesday, January 9. . . . Approximately 1,000 U.S. and Iraqi troops were in the area. At 7:00 a.m., the insurgents began firing on U.S. and Iraqi troops and their vehicles from sniper positions on the roofs and the doorways of buildings. They also coordinated their mortar fire, indicating the high degree of their training and cohesiveness. And they continued to fight, rather than running away from American forces, as the enemy typically had done, surprising American forces. . . . The insurgents occupied positions in successive buildings, and moved effectively from building to building as the American and Iraqi forces went from one to the next. The battalion from the Stryker Brigade called for close air support from Apache helicopters and F-18s, which targeted the snipers on the building roofs until roughly noon. On the ground, the U.S. battalion remained engaged for eleven hours.

This is only one example of ISIS engaging in warfare with soldiers. ISIS has fought in civil conflict for years, with many of those years racking up thousands or tens of thousands of deaths in battle—for example, 2,000 such deaths in 2004 and more than 3,600 in 2007 (Pettersson et al. 2019). However, ISIS was not just killing soldiers. According to the Global Terrorism Database, at the same time that ISIS was involved heavily on the battlefield it was also slaughtering civilians (LaFree and Dugan 2007). Killing civilians and killing soldiers are not mutually exclusive activities for violent organizations.

* * *

The previous examples raise the key question that this book examines: Why do insurgent organizations sometimes kill civilians? Some do so frequently, some do so occasionally, and some seem to never do so. Why? As we discuss later, recent research seeks to explain civilian victimization by insurgent groups, focusing on factors such as insurgent weakness or country regime type. In this book, we present a different explanation, building on what we refer to as *insurgent embeddedness*—the extent to which an insurgent group is enmeshed in relationships with the state, other insurgents, and the public. With this framework, we propose a set of hypotheses for why such a group might attack civilians. We introduce new data on insurgent groups and their attributes, and we empirically test our arguments. In addition to looking at civilian targeting, we also examine how insurgent embeddedness might affect terrorist attacks on schools, violence against the news media, and a proclivity for attacking the general public (not government officials or other symbolic types of targets). Finally, because of our focus on relationships, we present an in-depth explanation of insurgent alliances and rivalries, examining their changes and determinants over time. In total, the book provides a comprehensive look at how insurgent groups interact with other actors—and the implications for several types of bloodshed against civilians.

1.2. Key Concepts

Insurgent organizations are non-state actors that compete with the state in a civil conflict (Raleigh et al. 2010, 655). Although they are non-state actors, this does not necessarily mean they do not have the support of another state. Indeed, many insurgent groups receive some assistance, such as arms or safe haven from a country other than the one they are fighting. However, insurgent groups have some independence from state sponsors, some autonomy, and are not state actors such as the national military. We sometimes use the term militant group as a synonym for insurgent group, although we acknowledge that some authors view insurgent group as a more specific term and some view militant group as a broader term (Ezrow 2017; Thompson 2014). "Insurgent organizations" excludes pro-government militias (Carey, Mitchell, and Lowe 2013), which by definition are aligned with the state, not competing against it. We prefer the term insurgent *organization* because we are referring to formal organizations, but we also use the term "insurgent group" as a synonym.

The term insurgent organization overlaps with *terrorist organization*, which some studies define as any non-state political group that uses terrorism (Jones and Libicki 2008; Phillips 2015a), while others define it as a group that *primarily* uses terrorism (Cronin 2008). Because many or most insurgent groups use terrorism at some point, the former of these two definitions of terrorist group includes most insurgent groups, whereas the latter usually includes few insurgent groups. This is because for groups to compete with the state in civil conflict, there needs to be two-sided violence and a substantial number of deaths—and these two criteria are often not met by groups that *primarily* use terrorism.

The definition of the term *terrorism* is highly debated (Enders and Sandler 2011; Ortbals and Polini-Staudinger 2014; Weinberg, Pedahzur, and Hirsch-Hoefler 2004), but in the context of civil conflict, we define terrorism as when insurgent groups attack civilians—individuals who are not members of the government security forces (military or police). Focusing on civilians as victims, and excluding security forces, is important for differentiating terrorism from more traditional insurgent violence. We sometimes use the term noncombatants synonymously for civilians. Following other scholars of civil conflict (Fortna 2015, Stanton 2013, Thomas 2014), we use the terms terrorism, civilian targeting, and civilian victimization as synonyms in this context. Some definitions of terrorism refer to the motives of the perpetrators, such as desiring to spread fear in a wider audience (Enders and Sandler 2011). We assume such motives are present when insurgents are attacking civilians, but it is not possible to measure the actual motives of attackers. For this reason, we focus on the actual behavior of insurgent groups. Given the contested nature of the concept of terrorism, our empirical tests use different measures of terrorism.

Civil conflict derives from an incompatibility over control of the government or territory resulting in the use of armed force between the government and at least one non-state actor that leads to a substantial number of battle-related deaths. This definition draws on the "armed conflict" characterization from UCDP (Gleditsch et al. 2002; Pettersson et al. 2019). In this book, we use UCDP data for analysis and thus embrace its definition of what fits for an insurgent group in terms of the universe of groups we examine (Gleditsch et al. 2002; Pettersson et al. 2019). More details, such as minimum-death thresholds, are discussed in the empirical sections.

As noted previously, we describe insurgent embeddedness as the extent to which an insurgent group is enmeshed in relationships with the state, other insurgents, and the public. This concept emphasizes that insurgent groups interact with a bevy of actors, and these relationships have important consequences for conflict behavior. The concept of embeddedness has underpinned decades of research on social interactions, including work on social network analysis (Granovetter 1985; Polanyi 1944). We take as our starting point that it is not possible to understand the motives and operational choices of insurgent organizations without reference to the relational context in which they operate because relationships create and constrain opportunities for insurgencies.

As outlined in Chapter 2, three types of relationships are important when analyzing insurgent behavior: relationships with the state, peers, and the public. The state can build and change relationships with insurgencies. States may choose to ignore, mollify, or directly confront insurgents. Regarding peers, alliances with other insurgent organizations are often a key source of knowledge, information, technical expertise, and financial capital, but armed conflict with rival insurgents can also sap an organization of its capacity to act and ability to respond when state actors engage in violent efforts to destroy it. Regarding the public, local citizens can provide insurgents critical resources or at least a benign environment. Alternately, the public can become the eyes and ears of state security forces. These three types of relationships—state, peer, and public—interact to create the overall relational context in which insurgent organizations are embedded and in which they make choices about targeting.

The Moro Islamic Liberation Front (MILF) represents an excellent example of the complexity of insurgent embeddedness. Throughout its history, MILF has negotiated with the government, fought the government, cooperated with some militant groups, formed rivalries with other militant groups, and has provided social services for the community (Abuza 2005, Podder 2012, Taya 2007). Each of these actions created new opportunities to advance the organization's agenda but also created limits on its freedom to act. In this book, we focus on insurgent interactions with the state, other groups, and the public. However, because the state and the public are covered in a great deal of conflict research already, more of our empirical work focuses on insurgent intergroup alliances and rivalries (see Section III).

1.3. State of the Research Explaining Civilian Targeting by Insurgent Organizations

There is a huge literature on why groups turn to violence in the first place. There is a great deal of research about how discrimination, inequality, and oppression can lead people to pick up weapons (Fearon and Laitin 2003; Gurr 1970, 2000; Lichbach 1989, 1998). Gurr (1970, 2000) identifies oppression and discrimination as the key incentives for using violence. Lichbach

(1989, 1998) applies a collective action and rational actor perspective to why people would move to violence which is based on the balance between the dangers of collective dissent and the possible rewards. Fearon and Laitin (2003) argue for the importance of poverty and government ineffectiveness and weakness as key conditions that favor the outbreak of political violence by organizations. A substantial body of research tries to explain why people decide to rebel against governments (Collier 2000; Gluckman 2013; Petersen 2001; Weinstein 2006). There is also a great deal of literature on the use of terrorism (Bjørgo 2004; Crenshaw 1981; Enders and Sandler 2011; Ross 1993). Although there is much literature on both terrorism and insurgency, a key question that is missing from much of it is why insurgents would target civilians—something that many researchers define as terrorism (Asal et al. 2012; Stanton 2013).

Scholars have long sought to understand civilian victimization by states (Azam and Hoeffler 2002; Bjørgo 2004). Recently, a growing literature has focused on civilian targeting by insurgent groups—examining causes and consequences (Fortna 2015; Hinkkainen, Polo, and Reyes 2021; Thomas 2014). Regarding work that seeks to explain civilian targeting by insurgent groups—our main question—a variety of answers are proposed. Some explanations of civilian targeting by militant groups focus on state attributes, whereas others emphasize characteristics of the militant group. Some research explores relational aspects, but as we show, this area of the literature is underdeveloped.

Perhaps the most common type of state-based explanation for insurgent terrorism involves regime type (Eck and Hultman 2007; Hultman 2012; Keels and Kinney 2019; Stanton 2013). It is argued that militant groups in democratic countries are more likely to attack civilians because these regimes are especially sensitive to civilian losses—and more likely to capitulate in the face of such targeting. This latter argument is consistent with the notion that terrorism is more common in democratic countries, or at least partially democratic countries, than non-democratic countries (Chenoweth 2010; Gaibulloev, Piazza, and Sandler 2017). It is noteworthy that the majority of terrorism, according to some measures, occurs within the context of civil war (Findley and Young 2012).

Regarding explanations based on militant group dynamics, several authors argue that especially weak groups are more likely to attack civilians (Hultman 2007; Polo and Gleditsch 2016; Polo and González 2020; Wood 2010). This is familiar from the notion that terrorism is a weapon of the weak. Some studies

provide additional nuance to this argument, suggesting that internal fragmentation or lack of internal control explains civilian targeting (Abrahms
and Potter 2015; Humphries and Weinstein 2006). Wood, Kathman, and
Gent (2012) point to extractive capacity and defection deterrence, and given
the importance of these attributes to insurgent organizations, declining capability could provide groups with increased incentives to victimize civilians.
Wood (2014) argues for additional nuance based on the source of groups'
support. He finds that increased group capacity is associated with *less* civilian
targeting when groups rely on local support, but increased group capacity is
associated with *more* civilian targeting when groups use alternative sources
of support. This is consistent with other resource-based arguments of civilian
victimization (Fortna, Lotito, and Rubin 2018; Weinstein 2006).

Other arguments draw on territorial aspects of civil conflict to explain
terrorism within the conflict (Farrell, Findley, and Young 2019). Kalyvas
(2006) argues that the extent and nature of territorial control by insurgents
and the state explain civilian victimization. Fjelde and Hultman (2014) find
that civilian targeting is more likely—by insurgents and the government—in
areas inhabited by the enemy's ethnic constituency. Ash (2018) argues that
insurgents use terrorism to punish government supporters and destabilize
key regions.

In addition to these arguments, some studies incorporate relational elements, as we do. For example, Wood et al. (2012) argue that inter-insurgent violence leads to civilian targeting. This is related to Balcells' (2010) argument
that pre-war electoral violence leads to subsequent civilian victimization.
Salehyan, Siroky, and Wood (2014) raise the issue of principal–agent dynamics in civil conflict, and they note that foreign sponsorship of insurgents
could encourage or discourage civilian targeting. They find that insurgents
sponsored by democratic countries are less likely to attack civilians, whereas
insurgents sponsored by multiple countries are more likely to do so.

1.3.1. Shortcomings of the Civilian Targeting Literature

The previous discussion outlined explanations of civilian targeting by insurgent organizations, concluding by noting that only a few studies have
drawn on the actual features or behaviors of the organizations or the
relationships that insurgent organizations have with others as an explanation. However, it seems logical to think that the behavior, features of the

organization, and relationships would be important for behavior such as civilian targeting.

It is important to note that to date, most of the research has examined general state attributes such as regime type to proxy state actions—democracies are assumed to be more conciliatory, whereas autocracies are likely to be more coercive (e.g., Pape 1996; Stanton 2013). This approach is understandable given previous data shortcomings, but it is also problematic for at least two reasons. First, regime type is an imprecise indicator of specific state actions. Second, states in civil conflict often face multiple insurgent groups and can treat them quite differently. For example, the government of Colombia waged war against the National Liberation Army (ELN) while negotiating with the Revolutionary Armed Forces of Columbia (FARC) (BBC 2015).

A growing literature shows the importance of relational dynamics for militant groups. Regarding perhaps the most straightforward type of relations, interactions with other militant groups, scholars have explored important consequences of alliances and rivalry. Militant groups with allies tend to be more lethal (Asal and Rethemeyer 2008; Horowitz and Potter 2014), survive longer (Phillips 2014), and learn important tactics from each other (Horowitz 2010). Militants with rivals are often spurred to use new and horrific tactics (Bloom 2005; Conrad and Greene 2015; Conrad and Spaniel 2021; Metelits 2009), and competition-induced innovation can contribute to the durability of involved groups (Phillips 2015b). Researchers have also explored how ideology and the nature of the organization can have an impact on the organization's behavior (Asal and Rethemeyer 2008; Gutiérrez Sanin and Wood 2014; Horowitz and Potter 2014). Whereas there is much good research on violent organizational behavior, there is little on relational explanations for why some organizations attack civilians whereas others do not (Asal et al. 2019).

1.4. Overview of the Book

Building on an article written with Corina Simonelli and Joseph Young (Asal et al. 2019), this book contributes to addressing the question of why insurgent groups sometimes target civilians. We argue that civilian targeting is explained by what we refer to as insurgent embeddedness: insurgent organizations' relationships with the state, other insurgent groups, and the public. We introduce our data, the Big, Allied, and Dangerous II (BAAD2) Insurgency data (Asal and Rethemeyer 2015), on groups' attributes from 1998 to 2012. Four chapters provide batteries of empirical tests to show how insurgent attributes might be related to four dependent variables: civilian targeting in general, groups focusing the majority of their attacks on the general public, attacks on schools, and violence against journalists. In the final two analytical chapters, we offer in-depth analysis of interorganizational relations—alliances and rivalry. We conclude by summarizing and synthesizing our findings. We also offer suggestions for policymakers and indicate ways that other researchers can build on our theory and findings.

Specifically, Chapter 2 delves into the different theoretical perspectives that can be applied to help understand why insurgent organizations might kill civilians. One of the main foci is embeddedness theory, which argues that social interactions between entities will have a major impact on the behavior of organizations (e.g., see Granovetter 1985). Drawing on this perspective and other theoretical building blocks, we explore how the relationships between insurgent organizations and the state, other insurgents, and the public impact the likelihood that the organization will target civilians. This leads to a number of testable propositions. We theorize that government conciliatory measures—carrots—should lead to a lower likelihood of subsequent civilian targeting, whereas government coercion—sticks—should be associated with a greater chance of civilian targeting. As we turn to interorganizational relationships, we hypothesize that both insurgent alliances and rivalries should be associated with a greater propensity toward attacks on civilian. Regarding relationships between insurgents and the public, we argue that social service provision, ethnic motivations, and involvement in crime should lead to a greater likelihood of subsequent killing of civilians.

Chapter 3 presents the data that we use to test these hypotheses. To analyze the question of why insurgent organizations sometimes attack civilians, we use the BAAD2 data set, which is a yearly data set of insurgent organizations with data on their structure, behavior, and relations focusing on the years 1998–2012. The data include variables such as ideology, structure, leadership, social service provision, network connections to other insurgents, and levels and types of violence in each year. There are 140 organizations in the data set; these organizations were chosen from the UCDP conflict data (Gleditsch et al. 2002; Pettersson et al. 2019). By using UCDP, the data set codes identifiable organizations that have broken the 25 battle deaths threshold in at least one year and thus have crossed the line into what many scholars view as a defining characteristic of insurgents. If an organization passed the 25 battle deaths threshold, we coded it for the entire time period from 1998 to 2012 as long as they did not disband or stop using violence. The data suggest that there is a great deal of variation among insurgent groups regarding their use of civilian targeting. The rest of the chapter lays out the variables that are used in our analyses and how they are operationalized.

Chapter 4 analyzes the hypotheses about our key dependent variable: Why do insurgent organizations sometimes attack civilians? Drawing on the data presented in Chapter 3, we start with contingency tables, which suggest basic relationships between our main variables and attacking civilians. We use a variety of approaches to test the robustness of the findings related to the statistical analysis. Our primary model is a logistic regression in which the dependent variable is whether the organization attacked civilians in a particular year. We also use count models to explore the impact of our variables on the number of attacks against civilians, and also civilian attack fatalities, among other alternate dependent variables. We find support for our hypotheses, although some results are more robust than others. Government coercion (sticks), interorganizational alliances, interorganizational rivalry, and involvement in crime are consistently associated with civilian victimization or terrorism, measured multiple ways. Other results are mostly robust: Government concessions (carrots) are associated with less civilian targeting, and social service provision and ethnic motivations are associated with more civilian targeting. Overall, the insurgent embeddedness framework finds much support in the models.

Section II includes applications of embeddedness theory to three related outcomes: attacks on the general public, schools, and journalists. Chapter 5 explores why some insurgent groups *mostly* attack what we describe as the general public. In other words, we examine why some insurgent groups, during some years, focus their attacks on civilians who are not government employees or symbols of institutions such as religious leaders or members of the news media. This seemingly random type of targeting has been described as an "ideal type" of terrorism because it makes other members of the public fear that they might be the next victims. We find that government coercion, intergroup rivalries, and ethnic motivation seem to be important for explaining this type of targeting.

Chapter 6 focuses on the targeting of schools by insurgent organizations. We focus on this type of attack because it is truly one of the most heinous kinds of attack that a militant group can commit. Educational facilities are specifically named as protected entities in international law (International Peace Conference 1907). In addition, this type of attack has not been the subject of much systematic research. In this chapter, we examine again the key factors that we have identified as likely to be driving the attacking of civilians in general and specifically apply them to this kind of behavior. Attacks against schools are relatively rare, so perhaps unsurprisingly, we find that few variables are associated with such violence. No insurgent organization that received government concessions attacked an educational target in the subsequent year, consistent with expectations. On the other hand, government coercion is associated with insurgent organizations attacking schools. Two other factors linked to attacks on educational facilities are insurgent alliances and social service provision.

Chapter 7 examines another interesting and understudied outcome of interest: when insurgent organizations attack journalists. Violence against journalists is not simply violence against civilians; it is also a threat to communication and knowledge, including about governance and conflict. As with attacks against schools, given the relative rarity of such violence, it is somewhat difficult to identify explaining factors. It is also important to note that the variables that are related to violence against the news media are not all the same as in the previous analyses. With regard to violence against journalists, interorganizational alliances are one of the few variables that seem to help explain this type of bloodshed. This is perhaps due to organizations learning about tactics from their peers. In addition, insurgent groups in more democratic countries are more likely to attack the news media.

In Section III, Chapters 8 and 9 go in a different direction. Instead of examining determinants of violence against civilians, we provide more in-depth analysis of a key aspect of our embeddedness argument: interorganizational relationships. Specifically, we examine the factors that contribute to insurgents forming alliances or rivalries. In Chapter 8, we use analytically stochastic actor-oriented models to analyze the factors that impact the alliances that organizations form and keep with other insurgents. Analyses are conducted for theoretically relevant periods: the pre-September 11, 2001 (9/11) attack period (1998–2001), the post-9/11 time period (2002–2007), and what we call the post-surge period (2008–2012), as the United States removed most of its troops from Iraq. We find across all the time periods that overall connections are rare, and they vary by time period. Some organizations have numerous ties (e.g., al-Qaeda), but many organizations have none. Organizations are more likely to build ties with other organizations in the same country, as well as with organizations that are already allies of other allies of their own. In the first two time periods, organizations that were involved in terrorist attacks were more likely to make more ties. There are also several factors that have an impact for both periods after 9/11. For example, older organizations made fewer alliances. Overall, alliance formation is strongly affected by both temporal and organization-specific attributes.

In Chapter 9, we examine the factors that make organizations more likely to be rivals with each other. Unlike in Chapter 8, we are unable to use network-based models given the sparse nature of the number of rivals among insurgents. We thus report descriptive data to examine these rivalries, and we use logistic regressions of group-year data to determine what factors are related to insurgent organization rivalry. We find that ideology seems to be a key characteristic behind rivalry. Religious and separatist groups are likely to have rivals, but the type of ideology most likely to lead to rivalry is the combination of religious ethnoseparatist. Involvement in the drug trade and territorial control are also associated with the likelihood of rivalry. At the country level, perhaps surprisingly, democracy is negatively related to rivalry formation.

In Chapter 10, we summarize our key findings, including comparing results across the four main empirical chapters (Chapters 4–7). We discuss, for example, that inter-insurgent alliances are associated with attacks on civilians (Chapter 4), attacks on schools (Chapter 6), and attacks on journalists (Chapter 7), whereas intergroup rivalries are associated with attacks on civilians (Chapter 4) and attacks on the general public (Chapter 5). Possible reasons for these and other differences are considered. We also note important implications for policymakers and suggest future research building on the findings presented in this book.

References

Abrahms, Max, and Philip B. Potter. "Explaining Terrorism: Leadership Deficits and Militant Group Tactics." *International Organization* 69, no. 2 (2015): 311–42.
Abuza, Zachary. "The Moro Islamic Liberation Front at 20: State of the Revolution." *Studies in Conflict & Terrorism* 28, no. 6 (2005): 453–79.
Amaize, Emma, and Umar Yusuf. "Nigeria: Quit Order—Boko Haram Kills 30 Southerners." *All Africa* (2012). https://allafrica.com/stories/201201070005.html.
Asal, Victor, Luis De la Calle, Michael Findley, and Joseph Young. "Killing Civilians or Holding Territory? How to Think About Terrorism." *International Studies Review* 14, no. 3 (2012): 475–97.

Asal, Victor, Brian J. Phillips, R. Karl Rethemeyer, Corina Simonelli, and Joseph K. Young. "Carrots, Sticks, and Insurgent Targeting of Civilians." *Journal of Conflict Resolution* 63, no. 7 (2019): 1710–35.

Asal, Victor, and Karl Rethemeyer. "The Nature of the Beast: Organizational Structures and the Lethality of Terrorist Attacks." *Journal of Politics* 70, no. 2 (2008): 437–49.

Ash, Konstantin. "'The War Will Come to Your Street': Explaining Geographic Variation in Terrorism by Rebel Groups." *International Interactions* 44, no. 3 (2018): 411–36.

Azam, Jean-Paul, and Anke Hoeffler. "Violence Against Civilians in Civil Wars: Looting or Terror?" *Journal of Peace Research* 39, no. 4 (2002): 461–85.

Balcells, Laia. "Rivalry and Revenge: Violence Against Civilians in Conventional Civil Wars." *International Studies Quarterly* 54, no. 2 (2010): 291–313.

BBC. "Who Are Islamic Jihad?" June 9, 2003. http://news.bbc.co.uk/1/hi/world/middle_east/1658443.stm.

BBC. "Farc Rebels Say ELN Must Join Colombia Peace Process." May 13, 2015. Accessed July 19, 2020, from https://www.bbc.com/news/world-latin-america-32731754.

Bjørgo, Tore, ed. *Root Causes of Terrorism: Myths, Reality and Ways Forward.* New York: Routledge, 2004.

Bloom, Mia. *Dying to Kill: The Allure of Suicide Terror.* New York: Columbia University Press, 2005.

Carey, Sabine C., Neil J. Mitchell, and Will Lowe. "States, the Security Sector, and the Monopoly of Violence: A New Database on Pro-Government Militias." *Journal of Peace Research* 50, no. 2 (2013): 249–58.

Chenoweth, Erica. "Democratic Competition and Terrorist Activity." *Journal of Politics* 72, no. 1 (2010): 16–30.

Collier, Paul. "Rebellion as a Quasi-Criminal Activity." *Journal of Conflict Resolution* 44, no. 6 (2000): 839–53.

Conrad, Justin, and Kevin Greene. "Competition, Differentiation, and the Severity of Terrorist Attacks." *Journal of Politics* 77, no. 2 (2015): 546–61.

Conrad, Justin, and William Spaniel. *Militant Competition: How Terrorists and Insurgents Advertise with Violence and How They Can Be Stopped.* Cambridge University Press, 2021.

Council on Foreign Relations. "Backgrounder: Nigeria's Battle with Boko Haram." August 8, 2018. https://www.cfr.org/backgrounder/nigerias-battle-boko-haram.

Crenshaw, Martha. "The Causes of Terrorism." *Comparative Politics* 13, no. 4 (1981): 379–99.

Cronin, Audrey K. "How Terrorist Campaigns End." In *Leaving Terrorism Behind*, edited by Tore Bjorgo and John G. Horgan, 67–83. New York: Routledge, 2008.

Downes, Alexander B. "Restraint or Propellant? Democracy and Civilian Fatalities in Interstate Wars." *Journal of Conflict Resolution* 51, no. 6 (2007): 872–904.

Eck, Kistine, and Lisa Hultman. "One-Sided Violence Against Civilians in War: Insights from New Fatality Data." *Journal of Peace Research* 44, no. 2 (2007): 233–46.

Enders, Walter, and Todd Sandler. *The Political Economy of Terrorism.* Cambridge, UK: Cambridge University Press, 2011.

Ezrow, Natasha. *Global Politics and Violent Non-State Actors.* Thousand Oaks, CA: Sage, 2017.

Farrell, Megan M., Michael G. Findley, and Joseph Young. "Geographical Approaches in the Study of Terrorism." In *The Oxford Handbook of Terrorism*, edited by Erica

Chenoweth, Richard English, Andreas Gofas, and Stathis Kalyvas, 238–51. New York: Oxford University Press, 2019.

Fearon, James D., and David D. Laitin. "Ethnicity, Insurgency, and Civil War." *American Political Science Review* 97, no. 1 (2003): 75–90.

Findley, Michael G., and Joseph K. Young. "Terrorism and Civil War: A Spatial and Temporal Approach to a Conceptual Problem." *Perspectives on Politics* 10, no. 2 (2012): 285–305.

Fjelde, Hanna, and Lisa Hultman. "Weakening the Enemy: A Disaggregated Study of Violence Against Civilians in Africa." *Journal of Conflict Resolution* 58, no. 7 (2014): 1230–57.

Fortna, Virginia Page. "Do Terrorists Win? Rebels' Use of Terrorism and Civil War Outcomes." *International Organization* 69, no. 3 (2015): 519–56.

Fortna, Virginia Page, Nicholas J. Lotito, and Micheal A. Rubin. "Don't Bite the Hand That Feeds: Rebel Funding Sources and the Use of Terrorism in Civil Wars." *International Studies Quarterly* 62, no. 4 (2018): 782–94.

Freeman, Colin. "Nigeria Descends into Holy War; Christian Villagers in the North of the Country Live in Fear of the Boko Haram, an Islamist Sect Blamed in a String of Deadly Attacks." Canwest News Service. January 7, 2012. Accessed July 19, 2020, from https://advance.lexis.com/api/document?collection=news&id=urn:contentItem:54NM-VVS1-F125-13CW-00000-00&context=1516831.

Gaibulloev, Khusrav, James A. Piazza, and Todd Sandler. "Regime Types and Terrorism." *International Organization* 71, no. 3 (2017): 491–522.

Gleditsch, Niels P., Peter Wallensteen, Mikael Eriksson, Margareta Sollenberg, and Harvard Strand. "Armed Conflict 1946–2001: A New Dataset." *Journal of Peace Research* 39, no. 5 (2002): 615–37.

Gluckman, Max. *Order and Rebellion in Tribal Africa*. Vol. 4. New York: Routledge, 2013.

Granovetter, Mark. "Economic Action and Social Structure: The Problem of Embeddedness." *American Journal of Sociology* 91, no. 3 (1985): 481–510.

Gurr, Ted R. *Why Men Rebel*. Princeton, NJ: Princeton University Press, 1970.

Gurr, Ted R. *Peoples Versus States: Minorities at Risk in the New Century*. Washington, DC: US Institute of Peace Press, 2000.

Gutiérrez Sanín, Francisco, and Elisabeth Jean Wood. "Ideology in Civil War: Instrumental Adoption and Beyond." *Journal of Peace Research* 51, no. 2 (2014): 213–26.

Hinkkainen, Elliott Kaisa, Sara MT Polo, and Liana Eustacia Reyes. "Making Peace or Preventing It? UN Peacekeeping, Terrorism, and Civil War Negotiations." *International Studies Quarterly* 65, no. 1 (2021): 29–42.

Horowitz, Michael C. "Nonstate Actors and the Diffusion of Innovations: The Case of Suicide Terrorism." *International Organization* 64, no. 1 (2010): 33–64.

Horowitz, Michael C., and Philip B. K. Potter. "Allying to Kill: Terrorist Intergroup Cooperation and the Consequences for Lethality." *Journal of Conflict Resolution* 58, no. 2 (2014): 199–225.

Hultman, Lisa. "Battle Losses and Rebel Violence: Raising the Costs for Fighting." *Terrorism and Political Violence* 19, no. 2 (2007): 205–22.

Hultman, Lisa. "Attacks on Civilians in Civil War: Targeting the Achilles Heel of Democratic Governments." *International Interactions* 38, no. 2 (2012): 164–81.

International Peace Conference. "Convention (IV) Respecting the Laws and Customs of War on Land and Its Annex: Regulations Concerning the Laws and Customs of War on

Land." The Hague, the Netherlands, October 18, 1907. https://ihl-databases.icrc.org/ihl/INTRO/195.

Jones, Seth G., and Martin C. Libicki. *How Terrorist Groups End: Lessons for Countering Al Qa'ida*. Vol. 741. Santa Monica, CA: RAND, 2008.

Kagan, Kimberly. "Haifa Street: One Year Later" [online]. The Institute for the Study of War." 2008. Accessed June 6, 2020, from http://www.understandingwar.org/reference/haifa-street-one-year-later.

Kalyvas, Stathis N. *The Logic of Violence in Civil War*. Cambridge, UK: Cambridge University Press, 2006.

Keels, Eric, and Justin Kinney. "'Any Press Is Good Press?' Rebel Political Wings, Media Freedom, and Terrorism in Civil Wars." *International Interactions* 45, no. 1 (2019): 144–69.

LaFree, Gary, and Laura Dugan. "Introducing the Global Terrorism Database." *Terrorism and Political Violence* 19, no. 2 (2007): 181–204.

Lichbach, Mark I. "An Evaluation of 'Does Economic Inequality Breed Political Conflict?' Studies." *World Politics* 41, no. 4 (1989): 431–70.

Lichbach, Mark I. *The Rebel's Dilemma*. Ann Arbor, MI: University of Michigan Press, 1998.

Metelits, Claire M. "The consequences of rivalry: Explaining insurgent violence using fuzzy sets." *Political Research Quarterly* 62, no. 4 (2009): 673–84.

Mshelizza, Ibrahim. "Christians Flee Attacks in Northeast Nigeria." Reuters. January 7, 2012. https://uk.mobile.reuters.com/article/amp/idAFJOE80600P20120107.

Ortbals, Candice D., and Lori Poloni-Staudinger. "Women Defining Terrorism: Ethnonationalist, State, and Machista terrorism." *Critical Studies on Terrorism* 7, no.3 (2014): 336–56.

Osumah, Oarhe. "Boko Haram Insurgency in Northern Nigeria and the Vicious Cycle of Internal Insecurity." *Small Wars & Insurgencies* 24, no.3 (2013): 536–60.

Pape, Robert A. *Bombing to Win: Air Power and Coercion in War*. Ithaca, NY: Cornell University Press, 1996.

Petersen, Roger D. *Resistance and Rebellion: Lessons from Eastern Europe*. Cambridge, UK: Cambridge University Press, 2001.

Pettersson, Therese, Stina Högbladh, and Magnus Öberg. "Organized Violence, 1989–2018 and Peace Agreements." *Journal of Peace Research* 56, no. 4 (2019): 589–603.

Phillips, Brian J. "Terrorist Group Cooperation and Longevity." *International Studies Quarterly* 58, no.2 (2014): 336–47.

Phillips, Brian J. "What Is a Terrorist Group? Conceptual Issues and Empirical Implications." *Terrorism and Political Violence* 27, no.2 (2015a): 225–42.

Phillips, Brian J. "Enemies with Benefits? Violent Rivalry and Terrorist Group Longevity." *Journal of Peace Research* 52, no. 1 (2015b): 62–75.

Podder, Sukanya. "Legitimacy, Loyalty and Civilian Support for the Moro Islamic Liberation Front: Changing Dynamics in Mindanao, Philippines." *Politics, Religion & Ideology* 13, no.4 (2012): 495–512.

Polanyi, Karl. *The Great Transformation: Economic and Political Origins of Our Time*. New York: Rinehart, 1944.

Polo, Sara M. T., and Kristian Skrede Gleditsch. "Twisting Arms and Sending Messages: Terrorist Tactics in Civil War." *Journal of Peace Research* 53, no. 6 (2016): 815–29.

Polo, Sara M. T., and Belén González. "The Power to Resist: Mobilization and the Logic of Terrorist Attacks in Civil War." *Comparative Political Studies* 53, no. 13 (2020): 2029–60.

Raleigh, Clionadh, Andrew Linke, Havard Hegre, and Joakim Karlsen. "Introducing ACLED: An Armed Conflict Location and Event Dataset: Special Data Feature." *Journal of Peace Research* 47, no. 5 (2010): 651–60.

Roggio, Bill. "Task Force Warhorse: Classical Counterinsurgency on Haifa Street." *FDD's Long War Journal*. 2007. Accessed June 6, 2020, from https://www.longwarjournal.org/archives/2007/08/task_force_warhorse.php.

Ross, Jeffrey I. "Structural Causes of Oppositional Political Terrorism: Towards a Causal Model." *Journal of Peace Research* 30, no. 3 (1993): 317–29.

Salehyan, Idean, David Siroky, and Reed M. Wood. "External Rebel Sponsorship and Civilian Abuse: A Principal-Agent Analysis of Wartime Atrocities." *International Organization* 68, no. 3 (2014): 633–61.

Stanton, Jessica A. "Terrorism in the Context of Civil War." *Journal of Politics* 75, no. 4 (2013): 1009–22.

Taya, Shamsuddin L. "The Political Strategies of the Moro Islamic Liberation Front for Self-Determination in the Philippines." *Intellectual Discourse* 15, no. 1 (2007).

Thomas, Jakana. "Rewarding Bad Behavior: How Governments Respond to Terrorism in Civil War." *American Journal of Political Science* 58, no. 4 (2014): 804–18.

Thompson, Peter G. *Armed Groups: The 21st Century Threat*. Lanham, MD: Rowman & Littlefield, 2014.

Weinberg, Leonard, Ami Pedahzur, and Sivan Hirsch-Hoefler. "The Challenges of Conceptualizing Terrorism." *Terrorism and Political Violence* 16, no. 4 (2004): 777–94.

Weinstein, Jeremy M. *Inside Rebellion: The Politics of Insurgent Violence*. Cambridge, UK: Cambridge University Press, 2006.

Wood, Reed M. "Rebel Capability and Strategic Violence Against Civilians." *Journal of Peace Research* 47, no. 5 (2010): 601–14.

Wood, Reed M. "Opportunities to Kill or Incentives for Restraint? Rebel Capabilities, the Origins of Support, and Civilian Victimization in Civil War." *Conflict Management and Peace Science* 31, no. 5 (2014): 461–80.

Wood, Reed M., Jacob D. Kathman, and Stephen E. Gent. "Armed Intervention and Civilian Victimization in Intrastate Conflicts." *Journal of Peace Research* 49, no. 5 (2012): 647–60.

Zieve, Tamara. "This Week in History: Terror Attack on Bus 405." *Jerusalem Post*. July 1, 2012. Accessed July 19, 2020, from https://www.jpost.com/Features/In-Thespotlight/This-Week-In-History-Terror-attack-on-Bus-405.

2

The Embeddedness Theory of Civilian Targeting by Insurgent Organizations

2.1. Introduction

In 1999, the Revolutionary United Front, an insurgent organization in Sierra Leone, received international attention from the world's media—and not for good reasons (Human Rights Watch 1999). The Revolutionary United Front killed, raped, and maimed civilians (Human Rights Watch 1999):

> The rebels carried out large numbers of mutilations, in particular amputation of hands, arms, legs, and other parts of the body. In Freetown, several hundred people, mostly men but also women and children, were killed and maimed in this way. Twenty six civilians were the victims of double arm amputations. One eleven-year-old girl describes how she and two of her friends were taken away by a group of rebels, who then hacked off both of their hands.

This kind of slaughter and abuse of civilians by a non-state actor is the epitome of terrorism. The behavior of the Revolutionary United Front was horrific, extending to amputations and, according to some sources, cannibalism (McKay 2005). Its leader, Foday Sankoh, encouraged his soldiers to terrorize the population by enslaving children, rape, slaughter, and cannibalism, and his soldiers clearly followed through (*The Guardian* 2000). Sankoh and the Revolutionary United Front created bands of enslaved children, called Small Boy Units, which they taught to slaughter. In an interview with CNN (2001), one child related that

> when they caught us (my family) and took us to Koidu, I saw plenty of dead bodies and I was afraid. In fact, they almost killed my mother, saying, "Why

Insurgent Terrorism. Victor Asal, Brian J. Phillips, and R. Karl Rethemeyer, Oxford University Press. © University of Maryland National Consortium for the Study of Terrorism and Responses to Terrorism (START) 2022.
DOI: 10.1093/oso/9780197607015.003.0002

are you afraid?" Then they took me away from my mother. My mother was alone. So they took me then with my elder brother and trained us. They gave us gunpowder and we ate it. They put some other medicine in it and that gave us the mind to be able to kill.

After that, they caught somebody and gave me that person to kill, and I killed them. The blood, I also took it. They said I should rub it on my eyes, and I did it. That gave me the mind to kill.

Evil? Yes. Terrorist behavior? Yes. However, the Revolutionary United Front was not just a terrorist organization. From 1998—the year our data set starts—until 2000, the Revolutionary United Front was involved in more than 300 battle deaths per year: 719 battle deaths in 1998, 433 in 1999, and 355 in 2000. In 2001, it dropped to 48 deaths, and then the organization stopped killing soldiers and being killed by soldiers (Pettersson and Wallensteen 2015; Sundberg, Eck, and Kreutz 2012). As should be clear from what we described previously, however, the organization did not just kill soldiers. It killed 258 civilians in 1998, 18 in 1999, and 34 in 2000 (Asal, Rethemeyer, and Weedon 2015).

The Revolutionary United Front is just one example of an insurgent organization that spent its time killing not just soldiers but also civilians. Another example is the Al-Nusra Front involved in the Syrian civil war. In 2012, the Al-Nusra Front was involved in 339 battle deaths (Pettersson and Wallensteen 2015; Sundberg et al. 2012) but it also killed 132 civilians. Why would an insurgent organization such as the Revolutionary United Front and the insurgent organizations in Syria—and many others—kill civilians like this? All the organizations in the Uppsala Conflict Data Program data set, which is the basis of our analysis, are insurgent organizations, and all have been involved in at least 25 battle deaths in at least 1 year between 1998 and 2012. Many of these organizations have been involved in many more than 25 battle deaths in a year, and most have been involved in breaking the key bar of 25 battle deaths for many years (Pettersson and Wallensteen 2015; Sundberg et al. 2012). If an organization is successful at doing this, why would it turn to killing civilians? In this chapter, we outline our theory of civilian targeting, which we then test both in general and in a disaggregated manner in subsequent chapters.

2.2. Theory: Insurgent Embeddedness and Civilian Victimization

In our theoretical argument about why some organizations turn to killing civilians, the key underlying logic is that other actors with whom they might interact surround insurgent groups. They are embedded in dense networks of relationships. Insurgent groups engage with the state, other insurgent groups, the public, and other actors. However, not all insurgent groups interact with the same actors to the same degree. There is important variation in insurgent group interactions, and we argue that this variation helps explain why some insurgent groups are much more likely than others to attack civilians. Some research finds it parsimonious to downplay or ignore these relationships or only focus on one type of relationship—such as state sponsorship or rivalries with other insurgents. However, the emphasis of this book is that paying attention to this web of ties and potential ties is crucial for understanding group behavior.

2.2.1. Embeddedness Theory

Research programs across a number of disciplines emphasize the importance of social relations as fundamental for understanding individual and organizational behavior. Most visibly and explicitly oriented toward social interactions, social network analysis has developed over more than a century to help explain why actors (nodes) and their connections or ties matter (Wasserman and Faust 1994). Some view social network analysis as its own discipline or paradigm, whereas others view it as a methodology that scholars in many disciplines use. Regardless, this work became more formal and systematic in the latter half of the 20th century, with social scientists exploring topics from business to politics and crime (e.g., Eulau and Rothenberg 1986; Fraser and Hawkins 1984; Tichy, Tushman, and Fombrun 1979).

It is in this context that the concept of *embeddedness* emerged and developed. Polanyi (1944) introduced the term as a reaction to what he saw as atomistic views of human behavior in economics. He argued that the market was heavily embedded in social relations, including political and religious frameworks. Wrong (1961) described this view as "oversocialized," and other scholars apparently agreed. Most prominently, Granovetter (1985) rejected both classic liberalism as undersocialized and arguments such as Polanyi's for being too dependent on social roles. Granovetter argued for what could be called a Goldilocks-like just-right version of behavior that assumed human rationality and purposiveness but that actions are "embedded in concrete, ongoing systems of social relations" (p. 487). Granovetter's notion of embed-dedness motivates a great deal of research with social network assumptions, and it provides a crucial building block for our research.

Other social science research has benefitted from drawing on embed-dedness. For example, Ruggie's (1982) concept of embedded liberalism, the notion of economic liberalism enmeshed in the social community, drew on Polanyi's (1944) original concept. Scholars have addressed a crucial question in political science—why people vote when it might seem rational not to do so—by drawing on the embeddedness of voters in informal social networks (Abrams, Iversen, and Soskice 2011). Regarding organizations, scholars un-derstand U.S. corporate malfeasance by emphasizing firms' embeddedness in organizational and political structures (Prechel and Morris 2010). Firm embeddedness in relationships with other firms is associated with survival (Uzzi 1996). The study of unions has also benefitted from exploring their embeddedness in broader networks of other civic associations (Lee 2007). In international relations, scholars have examined networks of nations, exploring how the interactive nature of states has important implications (Maoz 2010; Maoz et al. 2006).

Some work on militant groups leverages embeddedness to understand subnational violence. Staniland (2014) argues that the social embeddedness of insurgent group leaders—the networks and institutions they were a part of before conflict began—are crucial for explaining insurgent group cohe-sion. Sarbahi (2014) explores how the extent of insurgent groups' embed-dedness in the population affects civil war outcomes. Other work examines how insurgent relationships have implications for a country long after a civil war has ended (Daly 2012; Sindre 2016a, 2016b; E. Wood 2008). A study of why some terrorist organizations use weapons of mass destruction finds it fruitful to consider groups' embeddedness in networks of other organiza-tions (Asal, Ackerman, and Rethemeyer 2012). Looking at related violent organizations, scholars of organized crime employ embeddedness theory to show that relationships among street gangs condition the extent to which government actions against them produce desired results (Vargas 2014). At the individual level, researchers have explored the extent to which criminals are socially embedded in their gangs (Pyrooz, Sweeten, and Piquero 2013; Smith and Papachristos 2016).

Without explicitly mentioning embeddedness, and focusing on various outcomes, other work on militant groups has drawn on the importance of groups' interactions with other actors, such as friendly and rivalrous organizations (Asal and Rethemeyer 2008) and state sponsors (Bapat 2012; Carter 2012). A growing line of research on terrorism and insurgency also draws on social network analysis, but the most frequent application is *within* groups or movements—for example, connections among individuals (Campedelli, Bartulovic, and Carley 2021; Everton 2012; Perliger and Pedahzur 2011; Raab and Milward 2003; Zech and Gabbay 2016). In the following sections, we draw on this and other research, including some of our own, to build an explanation of how insurgent embeddedness is crucial for explaining civilian targeting. Here, we outline three key aspects of insurgent embeddedness: interactions with the state that they primarily target, other insurgents, and the public.

2.2.2. Insurgent Embeddedness and Civil Conflict

In general, the notion of insurgent embeddedness emphasizes the fact that militants in civil conflict are enmeshed in complex relationships involving the state they fight, potentially other insurgent groups if there are multiple insurgent organizations, and the broader public. Figure 2.1 shows several ways that research on civil conflict incorporates different actors and their

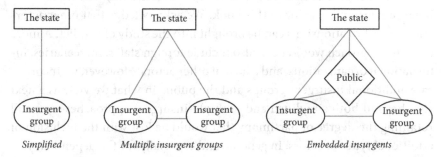

Figure 2.1 Three ways to characterize civil conflict. The gray lines indicate important relationships.

relationships. Many studies of civil conflict behavior, particularly those eval-
uated with quantitative tests, use a simplified model of conflict involving
"the state" against "the insurgents"—one general opposition category (e.g.,
Collier and Hoeffler 2004; Fearon and Laitin 2003). This approach black-
boxes insurgents into one category even when there might be multiple
insurgent organizations with complex relationships between them. This approach offers the advantage of parsimony, and it permits scholars to focus
on state attributes, but it also overlooks important dynamics. A second approach, depicted in the middle of Figure 2.1, acknowledges the possible existence of multiple militant organizations and relationships among them.
A growing line of research identifies important interactions involving insurgent groups, such as alliances and rivalries (Akcinaroglu 2012; Christia
2012; Fjelde and Nilsson 2012; Gade et al. 2019a; R. Wood and Kathman
2015). Some related work examines fragmentation of insurgent organizations (Bakke, Cunningham, and Seymour 2012; Mosinger 2018; Pearlman
and Cunningham 2012; Perkoski 2019). Overall, civil conflict research has
advanced substantially by considering interorganizational relationships.

The left-side diagram in Figure 2.1 shows a more complex depiction
of civil conflict. In this approach, intergroup insurgent dynamics are explicitly theorized, as are relationships between insurgents and the public
and between the state and the public. This approach takes into consideration the way that many insurgent groups are in fact embedded in a
complex series of relationships that affect them, other actors, and the conduct of conflict. As noted previously, the literature demonstrates critical
relationships between militant groups and the broader public—whether
via interpersonal political networks, social service provision, or otherwise
(Iannaccone and Berman 2006; Sarbahi 2014; Staniland 2014; E. Wood
2003). This degree of insurgent embeddedness is the way that we theorize
insurgent group behavior in this book. We acknowledge that other actors
and types of relationships can be brought into the study of conflict. A more
complex approach would be to also include foreign states, mercenaries, international organizations, and a host of other actors. However, we incorporate additional insurgent groups and the public in what we view as a next
step beyond both simplified and multiple-insurgent approaches—yet still
retaining the degree of parsimony that would be lost with the inclusion of
additional types of actors. In general, and especially for the dependent variable of this study, multiple insurgents and the public are the most important actors to consider alongside the state.

2.2.3. Insurgents and the Targeted State

Insurgent groups, by definition as discussed in Chapter 1, are engaged in battle with a state. Generally, this is the state in which they are primarily located, such as the Liberation Tigers of Tamil Eelam (LTTE) targeting the Sri Lankan government, the Revolutionary United Front attacking the government of Sierra Leone, and the Revolutionary Armed Forces of Colombia (FARC) seeking the overthrow of the Colombian government. Each state confronts "its" insurgent groups in unique ways. Here, drawing on Asal et al. (2019), we focus on two general approaches that states take toward dissident groups: conciliatory actions and coercive actions. These tactics are also referred to as carrots and sticks, respectively.

Governments frequently offer carrots to insurgent groups, from temporary ceasefires to negotiations and substantial political concessions (Bapat 2005; Duyvesteyn and Schuurman 2011). Sometimes these actions are unilateral, such as when Sudan unilaterally extended ceasefires with several insurgent groups in 2017 and 2018. Similarly, the Afghan government (along with U.S. partner forces) declared a unilateral ceasefire against the Taliban during Ramadan in 2018 (Lawrence and Garland 2018). Perhaps more often, concessions are part of an agreement, such as ceasefires between Myanmar and insurgent groups, in which the state stops coercive activity against the group in exchange for the group agreeing to do the same. The groups do not disarm, and they continue to train, but both sides agree not to attach each other.

State coercive actions are probably more familiar phenomena within the civil conflict environment. States bomb insurgent-held territory, arrest members and leaders, target leaders for assassination, and engage the group directly in combat. Some readers might think states and militant groups constantly fight, but there are many years during which militant groups are relatively dormant—perhaps because of state concessions or perhaps out of weakness. Similarly, states might not coerce insurgent groups every year because of lulls in group behavior or effective group evasion. Many states with insurgent groups in their territory are also weak, lacking substantial resources. Constant coercion can be costly, so these states face incentives to take breaks from counterinsurgency.

There is substantial variation across time and space regarding government tactics. At one point in the Sri Lankan civil war, insurgent group LTTE asked the government for a ceasefire to protect civilians. The government rejected the demand and continued to attack the group (Chamberlain 2009). States might shift between conciliation and coercion depending on relative strength or recent wins/losses, when a new government takes office, or out of other domestic politics concerns. These diverse actions should have divergent consequences for subsequent civilian targeting. Concessions are regularly offered specifically to prevent violence against civilians. For example, some research suggests that terrorist tactics lead to concessions (Saygili 2019; Thomas 2014). It is possible that this encourages more terrorism (Forsberg 2013; Walter 2004). This could be costly for leaders in terms of domestic opposition.

However, research that finds violence can induce concessions usually also finds that concessions to one group encourages violence by other groups (Forsberg 2013; Walter 2004). Our argument is different, looking at how concessions to one group affect subsequent violence by the same group. Although "biting the hand that feeds" is possible, it seems at least as likely that states rationally calculate and offer sufficient concessions to deter future "bad behavior." This would be consistent with evidence that the most common way that terrorist groups collapse (either ceasing to exist as an organization or giving up violence) is through political means, as opposed to military or law enforcement action (Jones and Libicki 2008). This suggests the following hypothesis:

H1: Concessions to insurgent groups are associated with a *lower* likelihood of civilian targeting.

Coercion should have the opposite effect. When states confront insurgent groups, the groups should respond in a variety of ways, including civilian targeting. Terrorism could be a choice—to inflict pain on populations that might be viewed as supportive of the state (Goodwin 2006). Alternately, it could be a last resort, as the groups have been weakened by state pressure. Either way, a number of studies find a backlash effect of state coercion, such as repression (Daxecker 2017; Polo 2020a, 2020b; Walsh and Piazza 2010). For example, Polo and Gleditsch (2016) find that countrywide repression increases the likelihood of subsequent terrorism, and it seems reasonable that actions targeted at specific groups would also increase the terrorism by those groups. Overall, coercion seems likely to lead to civilian deaths at the hands of insurgents. Thus, we hypothesize the following:

H2:   Coercion toward insurgent groups is associated with a *higher* likelihood of civilian targeting.

2.2.4. Insurgents and Other Insurgent Groups

Beyond relationships with states with which they bargain and battle, insurgent groups frequently interact with their own kind—other insurgent organizations. Insurgent groups are rarely alone in their conflict with the state.
In our data, the average insurgent organization exists in a country with two
other such groups. Some insurgents operate in states such as India with more
than a dozen groups, whereas others, such as those in Somalia and Iraq, co-
exist with at least five groups. Some groups are part of a pair in their state,
such as those in Liberia and Ethiopia. Beyond sharing space with other
groups in the same country, insurgents also frequently meet other militants
abroad—for example, in cross-border training camps. This makes sense
when one thinks about the many transnational dynamics of civil conflict,
with many insurgent groups crossing state lines for refuge (Salehyan 2011).
Intergroup dynamics have important implications for insurgent group behavior, including civilian targeting. In this section, we focus on two types of
interorganizational relationships long discussed in the literature: cooperation and rivalry. The former can be seen as positive embeddedness and the
latter as negative embeddedness, but they both are crucial aspects of insurgent groups enmeshed in webs of relationships with other actors.

Insurgent groups frequently cooperate, teaming up for training or joint
attacks, or even forming political coalitions (Bacon 2018; Moghadam 2017).
For example, several insurgent groups in Sudan organized a political front in
2011 to have more leverage in negotiations with the state (Small Arms Survey
2012). Nigeria's Boko Haram has cooperated with several other groups in the
region to train and carry out attacks, including the Movement for Oneness
and Jihad in West Africa, located in Mali (Zenn 2013). It is important to note
that cooperation is not permanent. In Colombia, for example, FARC fought
against its rival, the National Liberation Army, for years in the early 2000s but
announced a non-aggression pact in 2009. After the agreement, the organizations teamed up to attack the government (U.S. State Department 2010).
These relationships form for a number of reasons, from ideological affinity to
joint efforts to enhance capabilities against the state (Bapat and Bond 2012;
Blair et al. 2021; Bond 2010; Christia 2012; Gade et al. 2019a; Popovic 2018).

The state is generally the primary enemy of an insurgent group, but many insurgents also develop rivalries, sometimes violent, with other militant groups. Rivalry, most generally, is competition, and in many contexts multiple insurgent groups compete for potential members, donations, and other resources. This often occurs when groups seeking to represent the same community—for example, Palestinians or Northern Ireland Catholics—and thus contend for the same overall support base.[1] For instance, in the Philippines two groups compete to represent the Moro or Bangsamoro Muslim minority—the Moro Nationalist Liberation Front and the Moro Islamic Liberation Front. The latter group split off from the former, a dynamic of fragmentation that often occurs in civil conflict (Bakke et al. 2012). Although rivalry among insurgent groups is often nonviolent, and more about bidding for support, it sometimes involves direct attacks on each other. Groups in the Palestinian territories have often clashed violently, such as when Hamas members killed nine Fatah members in 2007 (Kershner 2007). Some scholars suggest that such violent rivalries stem from conflicts over material resources or political leverage (Berti 2020; Fjelde and Nilsson 2012). Other work has examined ideology and power asymmetry as roots of rivalry (Conrad, Greene, Phillips, and Daly 2021; Gade, Hafez, and Gabbay 2019b; Mendelsohn 2021; Phillips 2019; Pischedda 2020).

Each of these relationship types should contribute to increased civilian targeting. Cooperation is likely to affect civilian victimization because interorganizational alliances increase group capabilities and permit tactical diffusion. Regarding group capabilities, cooperation makes groups more equipped to carry out attacks, whether against civilians or military targets, due to the advantages of joint training and the larger numbers involved in joint attacks. The idea that stronger groups are more likely to attack civilians might seem counterintuitive given the notion of terrorism as a "weapon of the weak," but a growing literature suggests that this notion is incomplete. There is not much research on how alliances affect insurgent groups specifically, but several studies have examined the overlapping category of terrorist organizations and have found that alliances help these groups become more

[1] Rivalry among groups seeking to represent the same community has been referred to as intra-field rivalry (Phillips 2015). The literature also discusses inter-field rivalries—those between groups with opposite or at least substantially different goals, such as right wing versus left wing groups. However, this behavior is more common among terrorist groups, and it seems to occur less frequently in civil conflict. We do observe inter-field rivalries in civil conflict involving pro-government militias versus anti-regime groups, but pro-government militias are not considered in this book to be "insurgent groups," as is consistent with the civil conflict literature more broadly.

lethal and overall endure longer (Asal and Rethemeyer 2008; M. Horowitz and Potter 2012; Phillips 2014). Regarding insurgent groups in particular, Akcinaroglu (2012) finds that insurgents with strong and credible allies are likely to win the war.

Insurgent group cooperation also affects civilian targeting through tactical diffusion. Diffusion, including through learning, is an important phenomenon in political violence (Bara 2018). Militant group learning is highly important to their development and success, and some tactics learned are especially potent for attacking civilians.[2] M. Horowitz (2010) shows how intergroup alliances facilitated the spread of suicide bombing throughout militant groups. Other relevant tactics include the use of roadside bombs and kidnapping for ransom. These tactics have spread substantially in recent decades, with menacing implications for civilians. Overall, this suggests the following hypothesis:

H3: Intergroup alliances are associated with a *higher* likelihood of civilian targeting.

Rivalries are also likely to lead to increased civilian targeting by insurgent groups. This argument is perhaps more intuitive compared to the previous hypothesis. The literature suggests that rivalries are associated with more and extreme types of terrorism by militant groups (Belgioioso 2018; Bloom 2005; Conrad and Greene 2015; Dowd 2019; Nemeth 2014).[3] Bloom's (2005) argument focuses on outbidding; she argues that as groups compete for resources, they turn to more extreme tactics—such as suicide bombing—to draw attention to their organization over others. She argues that this is especially the case in societies in which such violence might be acceptable. Building on Bloom's work, Nemeth (2014) finds that intergroup competition is associated with increased terrorist violence.[4] Conrad and Greene (2015)

---

[2] Diffusion of tactics can *directly* lead to civilian targeting in this manner. Diffusion can also indirectly contribute to civilian targeting because the learning of new tactics also generally increases group capability, which in turn can facilitate attacks on civilians. We do not test which of these causal pathways is more important for connecting militant alliances to civilian targeting, but this would be a fruitful area for future research.

[3] Competition may lead not only to extreme violence but also can lead to ideological extremity (Walter 2017). Competition can also spur previously non-violent groups to start using violence (Asal and Phillips 2018).

[4] Nemeth's notion of competition only includes competition with groups of the same overall motivation type, such as religious groups competing with other religious groups and left-wing groups competing with other left-wing groups. He refers to this as "intraspecies competition," which is essentially the same as "intra-field rivalry" (Phillips 2015).

find that intergroup competition is associated with shocking and severe violence, including attacks on civilians. R. Wood and Kathman (2015) examine not only competition for resources but also direct intergroup violence. They find in a sample of African conflicts that violent rivalry is associated with increased civilian targeting. Metelits (2009) argues that a single-group monopoly can be beneficial to civilians, but "active rivalry" incentivizes noncombatant targeting. Note that the literature is not unanimous on rivalry leading to attacks on civilians. Findley and Young (2012), for example, find no relationship between militant group competition and either suicide terrorism or terrorism generally.[5] However, we find the logic to be compelling that competition should lead to attacks on civilians, and some evidence support this. Therefore, we propose the following hypothesis:

> H4: Intergroup rivalries are associated with a *higher* likelihood of civilian targeting.

2.2.5. Insurgents and the Public

In addition to their relationships with states and other militant organizations, insurgents are also embedded in dense relations involving the wider public (Arjona, Kasfir, and Mampilly 2015; Huber 2019; Stewart 2018; Tokdemir and Akcinaroglu 2016). This is the case for some groups more than others. Some insurgent organizations form, or have ex ante, tight relationships with local civilians, whereas others have more predatory and fleeting interactions. This is comparable to Olson's (1993) notion of stationary and roving bandits, where the former have incentives to protect civilians, whereas the latter do not. Within the counterinsurgency literature, scholars have long argued that conflict between insurgents and the state fundamentally depends on each side's relationship with the civilians in the middle. This is evident in the slew of clichés often used when analyzing civil conflict and the overlapping concept of insurgency. Mao said the guerrilla needs to travel among the people like a fish in the sea; winning the hearts and minds is essential. The 2006 U.S. military counterinsurgency manual argues that for counterinsurgents,

---

[5] Findley and Young (2012), like some other authors, measure competition by the number of militant groups in the country. While this is a way to measure general or potential competition, we measure rivalry in a more direct way, looking at whether each group is involved in a rivalry with another. This is discussed more in the next chapter.

the first step to follow once in an area is to "build trusted networks," which it explains is the "true meaning of the phrase 'hearts and minds'." (Petraeus and Amos 2006, A-5). Government forces need ties to the public to counteract those that insurgent groups likely already have. We focus on three observable indicators of insurgent–public interactions: social service provision, ethnic motivations, and involvement in crime.

Many insurgent groups seek to win over local populations by providing social services (e.g., Ibrahim Shire 2020). A growing literature seeks to understand this phenomenon of militant groups acting as quasi-states (Arjona et al. 2015; Florea 2020; Huang and Sullivan 2021; Stewart 2018). It creates a substantial challenge for authorities because if the populace no longer needs the government to provide services, the very state is obviated, ceding legitimacy to the insurgents (Grynkewich 2008).[6] There is a great deal of heterogeneity regarding how groups provide social services, whether creating their own nongovernment organizations or channeling their resources through extant nongovernment organizations (Asal, Flanigan, and Szekely 2020; Flanigan 2008; Parkinson 2013).

Social service provision affects conflict in diverse ways. Heger and Jung (2017) argue that service-providing groups are less likely to see spoilers break away during negotiation processes, which ultimately helps with peace talks. Insurgent groups that provide social services might receive more support from abroad (Flynn and Stewart 2018), and seem to last longer than other organizations (Wagstaff and Jung 2020).

A number of studies suggest a possible link between social service provision and civilian targeting. This might seem surprising because social services appear to indicate a concern about winning over popular support. So why might insurgents providing such services also attack civilians? One part of the answer is that the services might only be for a particular ethnic community. This set of civilians might be safe from civilian targeting, but others would not. A second dynamic is that insurgents deeply engaged with the populace might use both carrot and stick approaches themselves, thus both social services and (probably selective) civilian targeting. Regardless of the dynamics, insurgents engaged in social service provision should be especially equipped to effectively target civilians. Berman and co-authors draw on

---

[6] The implication for counterinsurgency is that social service provision by states can weaken the appeal of insurgent groups that had been offering such services (Grynkewich 2008). Generally consistent with this and arguments presented later, increased social welfare policies by governments are associated with less terrorism (Burgoon 2006).

the concept of club goods to argue that religious groups in particular can use social service distribution to increase the lethality of their violence (Berman and Laitin 2008; Iacconne and Berman 2006). Flanigan (2006) argues that social services permit groups to push communities toward greater acceptance of and participation in violence, including civilian targeting. These authors made their arguments using formal theory and drawing on a few cases with in-depth research, but questions remain about how generalizable the claims are. Thus, we hypothesize:

> H5: Social service provision is associated with a *higher* likelihood of civilian targeting.

Regardless of particular group activities such as social service provision, certain types of insurgent groups are likely to be deeply embedded within the broader community, which might in turn affect their propensity for civilian targeting. We focus on group motivations, which can range from "left" or "right" ideologically to purely religious or claiming to represent a particular ethnic community. Of these types of motivations, the latter category, ethnic motivations, should imply a relevant set of connections with the broader community beyond the insurgent group.

The notion of "ethnic group" is debated, but generally it is understood as a set of individuals who identify with each other because of a shared characteristic such as appearance, language, religion, or traditions (D. Horowitz 1985).[7] This suggests important bonds between members of an insurgent group motivated by ethnicity and all other members of the ethnicity. Ethnically motivated insurgent groups are more likely than other groups to rely on the community for public support. Weinstein (2006) discusses how groups with ethnic motivations are more likely to draw recruits from the broader ethnic community, and receive donations from the community as well. This suggests a different type of local embeddedness compared to groups that rely on drug production or state sponsors for funding.

Regarding how ethnic motivations might affect the propensity toward civilian targeting, the literature provides divergent expectations. Weinstein's (2006) work suggests that these groups' membership is more motivated to stay with the group, and their funding from the broader community is relatively stable. As a result, the groups do not need to rely as much on recruits

---

[7] See also Chandra (2006) for additional discussion.

who depend on a paycheck, nor do they have to rely on predatory practices such as extortion or forced recruiting. This suggests the recruits of ethnically motivated insurgent organizations are more disciplined and less likely to abuse civilians. This is consistent with work suggesting that ethnically motivated insurgents have advantages regarding recruiting and disciplining their members (Denny and Walter 2014; Gates 2002). In general, it might be reasonable to expect that groups tied to the local community—dependent on it—should be more likely to be protective of it and refrain from civilian targeting. In addition, some literature suggests that religious, not ethnically motivated, militant groups are the most violent (Iannaccone and Berman 2006). Taken together, this suggests that ethnic groups *should not* be more likely than other group types to target civilians.

On the other hand, there are some compelling reasons to expect that groups with ethnic motivations should be *more* likely to attack civilians. Militant groups seeking to represent an ethnic community frequently fragment, and this results in infighting and substantial co-ethnic civilian casualties (Ahmad 2016; Cunningham, Bakke, and Seymour 2012). Examples include the Tamil groups of Sri Lanka in the 1980s and Palestinian groups. Recent research suggests that the types of militant groups most likely to attack each other are those claiming to represent the same ethnic group (Conrad, Greene, Phillips, and Daly 2021). Civilians often get caught in the middle.

In addition, ethnically motivated insurgents often face rivals representing another ethnic community, as was the case with Hutu and Tutsi groups in the Great Lakes region of eastern Africa.[8] This, too, can lead to heavy civilian casualties. Groups can punish their rival by attacking any civilians from the rival's ethnic group. This can lead to spiraling tit-for-tat violence, made easier by the fact that civilians are soft targets, and members of particular ethnic groups are often more identifiable than, for example, leftists or government supporters. The security dilemma is especially strong with sectarian conflict (Posen 1993). Overall, we find it more likely that insurgent groups with ethnic motivations should be more likely to target civilians:

H6: Ethnic motivations are associated with a *higher* likelihood of civilian targeting.

---

[8] For example, in the Democratic Republic of Congo, the primarily Hutu Democratic Forces for the Liberation of Rwanda fought against the Tutsi-sympathetic National Congress for the Defense of the People.

Finally, insurgent groups interact with the public due to these groups' involvement in criminal activities. Many insurgents demand extortion from citizens, and others raise funds by trafficking drugs or other contraband (Asal, Milward, and Schoon 2015). Still others kidnap or directly rob citizens. Insurgents substantially involved in criminal behavior include FARC, Abu Sayyaf, and the Tamil Tigers (e.g., Gutiérrez Sanín 2004; Makarenko 2003; O'Brien 2012). There is a great deal of variety among insurgent organizations, however, because many do not engage in criminal activities. It seems likely that involvement in crime should be related to the use terrorism by insurgents.

Insurgent organizations substantially involved in crime tend to have predatory relationships with regular citizens, extracting resources coercively from them to fund the insurgency. This type of dynamic suggests that the insurgents are not overly concerned with having legitimacy in the eyes of the public or winning local hearts and minds. This is not to say that crime "causes" groups to use terrorism but, rather, that the same underlying factors might lead groups to engage in both types of behavior. In addition, once a group is engaging in one type of behavior, it might have fewer incentives to refrain from the other behavior. Crime, like violence against civilians, gives insurgents a bad reputation (Tokdemir and Akcinaroglu 2016). Groups without tight social ties to the local public often need to rely on economic incentives to recruit members, and this can draw opportunistic recruits who are more likely to abuse the public (Weinstein 2006). Overall, insurgents' relationships with the public matter, and involvement in serious crime suggests a predatory relationship with the public and therefore a greater chance of attacks on the public in terms of terrorism.

H7: Involvement in crime is associated with a *higher* likelihood of civilian targeting.

2.3. Conclusion

This chapter has outlined our theory of insurgent embeddedness and civilian targeting. Previous explanations of terrorism by insurgent groups have focused on group weakness, regime type, and other attributes that are parsimonious but downplay the complex set of relationships in which insurgent organizations are involved. Groups interact with the state, other militant organizations, and the broader public. These forms of embeddedness have important implications for civilian targeting. We focused on seven relationship aspects or indicators: government concession, government coercion, intergroup alliances, intergroup rivalry, social service provision, ethnic motivations, and crime. Each should be associated with a group's likelihood of attacking civilians. Chapter 3 introduces the data and Chapter 4 subjects the hypotheses to empirical testing.

References

Abrams, Samuel, Torben Iversen, and David Soskice. "Informal Social Networks and Rational Voting." *British Journal of Political Science* 41, no. 2 (2011): 229–57.

Ahmad, Aisha. "Going Global: Islamist Competition in Contemporary Civil Wars." *Security Studies* 25, no. 2 (2016): 353–84.

Akcinaroglu, Seden. "Rebel Interdependencies and Civil War Outcomes." *Journal of Conflict Resolution* 56, no. 5 (2012): 879–903.

Arjona, Ana, Nelson Kasfir, and Zachariah Mampilly, eds. *Rebel Governance in Civil War.* Cambridge, UK: Cambridge University Press, 2015.

Asal, Victor, H. Brinton Milward, and Eric W. Schoon. "When Terrorists Go Bad: Analyzing Terrorist Organizations' Involvement in Drug Smuggling." *International Studies Quarterly* 59, no. 1 (2015): 112–23.

Asal, Victor, and Brian J. Phillips. "What Explains Ethnic Organizational Violence? Evidence from Eastern Europe and Russia." *Conflict Management and Peace Science* 35, no. 2 (2018): 111–31.

Asal, Victor, Brian J. Phillips, R. Karl Rethemeyer, Corina Simonelli, and Joseph K. Young. "Carrots, Sticks, and Insurgent Targeting of Civilians." *Journal of Conflict Resolution* 63, no. 7 (2019): 1710–35.

Asal, Victor, and R. Karl Rethemeyer. "The Nature of the Beast: Organizational Structures and the Lethality of Terrorist Attacks." *Journal of Politics* 70, no. 2 (2008): 437–49.

Asal, Victor H., R. Karl Rethemeyer, and Suzanne Weedon. *Big Allied and Dangerous (BAAD) Codebook.* Albany, NY: Project on Violent Conflict, 2015.

Asal, Victor H., Gary A. Ackerman, and R. Karl Rethemeyer. "Connections Can Be Toxic: Terrorist Organizational Factors and the Pursuit of CBRN Weapons." *Studies in Conflict & Terrorism* 35, no. 3 (2012): 229–54.

Asal, Victor, Shawn Flanigan, and Ora Szekely. "Doing Good While Killing: Why Some Insurgent Groups Provide Community Services." *Terrorism and Political Violence* (2020): Ahead of print. https://doi.org/10.1080/09546553.2020.1745775

Bacon, Tricia. *Why Terrorist Groups Form International Alliances.* Philadelphia, PA: University of Pennsylvania Press, 2018.

Bakke, Kristin M., Kathleen Gallagher Cunningham, and Lee J. M. Seymour. "A Plague of Initials: Fragmentation, Cohesion, and Infighting in Civil Wars." *Perspectives on Politics* 10, no. 2 (2012): 265–83.

Bapat, Navin A. "Insurgency and the Opening of Peace Processes." *Journal of Peace Research* 42, no. 6 (2005): 699–717.

Bapat, Navin A. "Understanding State Sponsorship of Militant Groups." *British Journal of Political Science* 42, no. 1 (2012): 1–29.

Bapat, Navin A., and Kanisha D. Bond. "Alliances Between Militant Groups." *British Journal of Political Science* 42, no. 4 (2012): 793–824.

Bara, Corinne. "Legacies of Violence: Conflict-Specific Capital and the Postconflict Diffusion of Civil War." *Journal of Conflict Resolution* 62, no. 9 (2018): 1991–2016.

Belgioioso, Margherita. "Going Underground: Resort to Terrorism in Mass Mobilization Dissident Campaigns." *Journal of Peace Research* 55, no. 5 (2018): 641–55.

Berman, Eli, and David D. Laitin. "Religion, Terrorism and Public Goods: Testing the Club Model." *Journal of Public Economics* 92, nos. 10–11 (2008): 1942–67.

Berti, Benedetta. "From Cooperation to Competition: Localization, Militarization and Rebel Co-Governance Arrangements in Syria." *Studies in Conflict & Terrorism* (2020): 1–19.

Blair, C. W., E. Chenoweth, M. C. Horowitz, E. Perkoski, and P. B. Potter. "Honor Among Thieves: Understanding Rhetorical and Material Cooperation Among Violent Nonstate Actors." *International Organization* (2021): 1–40. doi:10.1017/S0020818321000114.

Bloom, Mia. *Dying to Kill: The Allure of Suicide Terror*. New York, NY: Columbia University Press, 2005.

Bond, Kanisha D. "Power, Identity, Credibility & Cooperation: Examining the Development of Cooperative Arrangements Among Violent Non-State Actors." Doctoral dissertation, The Pennsylvania State University, 2010. https://etda.libraries. psu.edu/catalog/11071.

Burgoon, Brian. "On Welfare and Terror: Social Welfare Policies and Political–Economic Roots of Terrorism." *Journal of Conflict Resolution* 50, no. 2 (2006): 176–203.

Campedelli, Gian Maria, Mihovil Bartulovic, and Kathleen M. Carley. "Learning Future Terrorist Targets Through Temporal Meta-Graphs." *Scientific Reports* 11, no. 1 (2021): 1–5.

Carter, David B. "A Blessing or a Curse? State Support for Terrorist Groups." *International Organization* 66, no. 1 (2012): 129–51.

Chamberlain, Gethin. "Sri Lanka Rejects Tamil Tigers' Ceasefire." *The Guardian*, April 27, 2009). Retrieved May 8, 2019, from https://www.theguardian.com/world/2009/apr/ 26/sri-lanka-tamil-tigers-ceasefire.

Chandra, Kanchan. "What Is Ethnic Identity and Does It Matter?" *Annual Review of Political Science* 9 (2006): 397–424.

Christia, Fotini. *Alliance Formation in Civil Wars*. Cambridge, UK: Cambridge University Press, 2012.

CNN. "CNN.com—Transcripts." 2001. http://transcripts.cnn.com/TRANSCRIPTS/ 0112/23/cp.00.html.

Collier, Paul, and Anke Hoeffler. "Greed and Grievance in Civil War." *Oxford Economic Papers* 56, no. 4 (2004): 563–95.

Conrad, Justin, and Kevin Greene. "Competition, Differentiation, and the Severity of Terrorist Attacks." *Journal of Politics* 77, no. 2 (2015): 546–61.

Conrad, Justin, Kevin T. Greene, Brian J. Phillips, and Samantha Daly. "Competition from Within: Ethnicity, Power, and Militant Group Rivalry." *Defence and Peace Economics* (2021): Ahead of print. doi:10.1080/10242694.2021.1951595

Daly, Sarah Zukerman. "Organizational Legacies of Violence: Conditions Favoring Insurgency Inset in Colombia, 1964–1984." *Journal of Peace Research* 49, no. 3 (2012): 473–91.

Daxecker, Ursula. "Dirty Hands: Government Torture and Terrorism." *Journal of Conflict Resolution* 61, no. 6 (2017): 1261–89.

Denny, Elaine K., and Barbara F. Walter. "Ethnicity and Civil War." *Journal of Peace Research* 51, no. 2 (2014): 199–212.

Dowd, Caitriona. "Fragmentation, Conflict, and Competition: Islamist Anti-Civilian Violence in Sub-Saharan Africa." *Terrorism and Political Violence* 31, no. 3 (2019): 433–53.

Duyvesteyn, Isabelle, and Bart Schuurman. "The Paradoxes of Negotiating with Terrorist and Insurgent Organisations." *Journal of Imperial and Commonwealth History* 39, no. 4 (2011): 677–92.

Eulau, Heinz, and Lawrence Rothenberg. "Life Space and Social Networks as Political Contexts." *Political Behavior* 8, no. 2 (1986): 130–57.

Everton, Sean F. *Disrupting Dark Networks* (no. 34). Cambridge, UK: Cambridge University Press, 2012.

Fearon, James D., and David D. Laitin. "Ethnicity, Insurgency, and Civil War." *American Political Science Review* 97, no. 1 (2003): 75–90.

Findley, Michael G., and Joseph K. Young. "More Combatant Groups, More Terror? Empirical Tests of an Outbidding Logic." *Terrorism and Political Violence* 24, no. 5 (2012): 706–21.

Fjelde, Hanne, and Desirée Nilsson. "Rebels Against Rebels: Explaining Violence Between Rebel Groups." *Journal of Conflict Resolution* 56, no. 4 (2012): 604–28.

Flanigan, Shawn Teresa. "Charity as Resistance: Connections Between Charity, Contentious Politics, and Terror." *Studies in Conflict & Terrorism* 29, no. 7 (2006): 641–55.

Flanigan, Shawn Teresa. "Nonprofit Service Provision by Insurgent Organizations: The Cases of Hizballah and the Tamil Tigers." *Studies in Conflict & Terrorism* 31, no. 6 (2008): 499–519.

Florea, Adrian. "Rebel Governance in de facto States." *European Journal of International Relations* 26, no. 4 (2020): 1004–31.

Flynn, D. J., and Megan A. Stewart. "Secessionist Social Services Reduce the Public Costs of Civilian Killings: Experimental Evidence from the United States and the United Kingdom." *Research & Politics* 5, no. 4 (2018): 1–10.

Forsberg, Erika. "Do Ethnic Dominoes Fall? Evaluating Domino Effects of Granting Territorial Concessions to Separatist Groups." *International Studies Quarterly* 57, no. 2 (2013): 329–40.

Fraser, Mark, and J. David Hawkins. "Social Network Analysis and Drug Misuse." *Social Service Review* 58, no. 1 (1984): 81–97.

Gade, Emily Kalah, Michael Gabbay, Mohammed M. Hafez, and Zane Kelly. "Networks of Cooperation: Rebel Alliances in Fragmented Civil Wars." *Journal of Conflict Resolution* 63, no. 9 (2019a): 2071–97.

Gade, Emily Kalah, Mohammed M. Hafez, and Michael Gabbay. "Fratricide in Rebel Movements: A Network Analysis of Syrian Militant Infighting." *Journal of Peace Research* 56, no. 3 (2019b): 321–35.

Gallagher Cunningham, Kathleen, Kristin M. Bakke, and Lee J. M. Seymour. "Shirts Today, Skins Tomorrow: Dual Contests and the Effects of Fragmentation in Self-Determination Disputes." *Journal of Conflict Resolution* 56, no. 1 (2012): 67–93.

Gates, Scott. "Recruitment and Allegiance: The Microfoundations of Rebellion." *Journal of Conflict Resolution* 46, no. 1 (2002): 111–30.

Goodwin, Jeff. "A Theory of Categorical Terrorism." *Social Forces* 84, no. 4 (2006): 2027–46.

Granovetter, Mark. "Economic Action and Social Structure: The Problem of Embededness." *American Journal of Sociology* 91, no. 3 (1985): 481–510.

Grynkewich, Alexus G. "Welfare as Warfare: How Violent Non-State Groups Use Social Services to Attack the State." *Studies in Conflict & Terrorism* 31, no. 4 (2008): 350–70.

Gutiérrez Sanín, Francisco. "Criminal Rebels? A Discussion of Civil War and Criminality from the Colombian Experience." *Politics & Society* 32, no. 2 (2004): 257–85.

Heger, Lindsay L., and Danielle F. Jung. "Negotiating with Rebels: The Effect of Rebel Service Provision on Conflict Negotiations." *Journal of Conflict Resolution* 61, no. 6 (2017): 1203–29.

Horowitz, Donald L. *Ethnic Groups in Conflict*. Berkeley, CA: University of California Press, 1985.

Horowitz, Michael C. "Nonstate Actors and the Diffusion of Innovations: The Case of Suicide Terrorism." *International Organization* 64, no. 1 (2010): 33–64.

Huang, Reyko, and Patricia L. Sullivan. "Arms for Education? External Support and Rebel Social Services." *Journal of Peace Research* 58, no. 4 (2021): 794–808.

Huber, Laura. "When Civilians Are Attacked: Gender Equality and Terrorist Targeting." *Journal of Conflict Resolution* 63, no. 10 (2019): 2289–2318.

Human Rights Watch. "Shocking War Crimes in Sierra Leone." 1999. https://www.hrw.org/news/1999/06/24/shocking-war-crimes-sierra-leone.

Iannaccone, Laurence R., and Eli Berman. "Religious Extremism: The Good, the Bad, and the Deadly." *Public Choice* 128, nos. 1–2 (2006): 109–29.

Ibrahim Shire, Mohammed. "More Attacks or More Services? Insurgent Groups' Behaviour During the COVID-19 Pandemic in Afghanistan, Syria, and Somalia." *Behavioral Sciences of Terrorism and Political Aggression* (2020): Ahead of print. https://doi.org/10.1080/19434472.2020.1856911

Kershner, Isabel. "Hamas Attacks Against Fatah Kill 14 and Add to Gaza Chaos." *New York Times*, May 16, 2007. Retrieved May 16, 2019, from https://www.nytimes.com/2007/05/16/world/middleeast/16gaza.html.

Jones, Seth G., and Martin C. Libicki. *How Terrorist Groups End: Lessons for Countering Al Qa'ida*. Santa Monica, CA: Rand Corporation, 2008.

Lawrence, J. P., and Chad Garland. "Afghan President, US Forces Commit to Weeklong Unilateral Cease-Fire with Taliban." *Stars and Stripes*, June 7, 2018. Retrieved May 8, 2019, from https://www.stripes.com/news/middle-east/afghan-president-us-forces-commit-to-weeklong-unilateral-cease-fire-with-taliban-1.531511.

Lee, Cheol-Sung. "Labor Unions and Good Governance: A Cross-National, Comparative Analysis." *American Sociological Review* 72, no. 4 (2007): 585–609.

Makarenko, Tamara. "'The Ties That Bind': Uncovering the Relationship Between Organised Crime and Terrorism." *Global Organized Crime* (2003): 159–92.

Maoz, Zeev. "Networks of Nations: The Evolution, Structure, and Impact of International Networks, 1816–2001" (Vol. 32). Cambridge, UK: Cambridge University Press, 2010.

Maoz, Zeev, Ranan D. Kuperman, Lesley Terris, and Ilan Talmud. "Structural Equivalence and International Conflict: A Social Networks Analysis." *Journal of Conflict Resolution* 50, no. 5 (2006): 664–89.

McKay, Susan. "Girls as 'Weapons of Terror' in Northern Uganda and Sierra Leonean Rebel Fighting Forces." *Studies in Conflict & Terrorism* 28, no. 5 (2005): 385–97.

Mendelsohn, Barak. "The Battle for Algeria: Explaining Fratricide Among Armed Nonstate Actors." *Studies in Conflict & Terrorism* 44, no. 9 (2021): 776–98.

Metelits, Claire. *Inside Insurgency: Violence, Civilians, and Revolutionary Group Behavior.* New York, NY: New York University Press, 2009.

Moghadam, Assaf. *Nexus of Global Jihad: Understanding Cooperation Among Terrorist Actors.* New York, NY: Columbia University Press, 2017.

Mosinger, Eric S. "Brothers or Others in Arms? Civilian Constituencies and Rebel Fragmentation in Civil War." *Journal of Peace Research* 55, no. 1 (2018): 62–77.

Nemeth, Stephen. "The Effect of Competition on Terrorist Group Operations." *Journal of Conflict Resolution* 58, no. 2 (2014): 336–62.

O'Brien, McKenzie. "Fluctuations Between Crime and Terror: The Case of Abu Sayyaf's Kidnapping Activities." *Terrorism and Political Violence* 24, no. 2 (2012): 320–36.

Olson, Mancur. "Dictatorship, Democracy, and Development." *American Political Science Review* 87, no. 3 (1993): 567–76.

Parkinson, Sarah Elizabeth. "Organizing Rebellion: Rethinking High-Risk Mobilization and Social Networks in War." *American Political Science Review* 107, no. 3 (2013): 418–32.

Pearlman, Wendy, and Kathleen Gallagher Cunningham. "Nonstate Actors, Fragmentation, and Conflict Processes." *Journal of Conflict Resolution* 56, no. 1 (2012): 3–15.

Perkoski, Evan. "Internal Politics and the Fragmentation of Armed Groups." *International Studies Quarterly* 63, no. 4 (2019): 876–89.

Perliger, Arie, and Ami Pedahzur. "Social Network Analysis in the Study of Terrorism and Political Violence." *PS: Political Science and Politics* 44, no. 1 (2011): 45–50.

Petraeus, David H., and James F. Amos. "Counterinsurgency: FM 3-24." *Headquarters Department of the Army* (2006).

Pettersson, Thérèse, and Peter Wallensteen. "Armed Conflicts, 1946–2014." *Journal of Peace Research* 52, no. 4 (2015): 536–50.

Phillips, Brian J. "Terrorist Group Cooperation and Longevity." *International Studies Quarterly* 58, no. 2 (2014): 336–47.

Phillips, Brian J. "Enemies with Benefits? Violent Rivalry and Terrorist Group Longevity." *Journal of Peace Research* 52, no. 1 (2015): 62–75.

Phillips, Brian J. "Terrorist Group Rivalries and Alliances: Testing Competing Explanations." *Studies in Conflict & Terrorism* 42, no. 11 (2019): 997–1019.

Pischedda, Costantino. *Conflict Among Rebels: Why Insurgent Groups Fight Each Other.* New York, NY: Columbia University Press, 2020.

Polanyi, K. *The Great Transformation: Economic and Political Origins of Our Time.* New York, NY: Rinehart, 1944.

Polo, Sara M. T. "How Terrorism Spreads: Emulation and the Diffusion of Ethnic and Ethnoreligious Terrorism." *Journal of Conflict Resolution* 64, no. 10 (2020a): 1916–42.

Polo, Sara M. T. "The Quality of Terrorist Violence: Explaining the Logic of Terrorist Target Choice." *Journal of Peace Research* 57, no. 2 (2020b): 235–50.

Polo, Sara M. T., and Kristian Skrede Gleditsch. "Twisting Arms and Sending Messages: Terrorist Tactics in Civil War." *Journal of Peace Research* 53, no. 6 (2016): 815–29.

Popovic, Milos. "Inter-Rebel Alliances in the Shadow of Foreign Sponsors." *International Interactions* 44, no. 4 (2018): 749–76.

Posen, Barry R. "The Security Dilemma and Ethnic Conflict." *Survival* 35, no. 1 (1993): 27–47.

Prechel, Harland, and Theresa Morris. "The Effects of Organizational and Political Embeddedness on Financial Malfeasance in the Largest US Corporations: Dependence, Incentives, and Opportunities." *American Sociological Review* 75, no. 3 (2010): 331–54.

Pyrooz, David C., Gary Sweeten, and Alex R. Piquero. "Continuity and Change in Gang Membership and Gang Embeddedness." *Journal of Research in Crime and Delinquency* 50, no. 2 (2013): 239–71.

Raab, Jörg, and H. Brinton Milward. "Dark Networks as Problems." *Journal of Public Administration Research and Theory* 13, no. 4 (2003): 413–39.

Ruggie, John Gerard. "International Regimes, Transactions, and Change: Embedded Liberalism in the Postwar Economic Order." *International Organization* 36, no. 2 (1982): 379–415.

Salehyan, Idean. *Rebels Without Borders: Transnational Insurgencies in World Politics.* Ithaca, NY: Cornell University Press, 2011.

Sarbahi, Anoop K. "Insurgent-Population Ties and the Variation in the Trajectory of Peripheral Civil Wars." *Comparative Political Studies* 47, no. 10 (2014): 1470–1500.

Saygili, Aslihan. "Concessions or Crackdown: How Regime Stability Shapes Democratic Responses to Hostage Taking Terrorism." *Journal of Conflict Resolution* 63, no. 2 (2019): 468–501.

Sindre, Gyda Marås. "In Whose Interests? Former Rebel Parties and Ex-Combatant Interest Group Mobilisation in Aceh and East Timor." *Civil Wars* 18, no. 2 (2016a): 192–213.

Sindre, Gyda Marås. "Internal Party Democracy in Former Rebel Parties." *Party Politics* 22, no. 4 (2016b): 501–11.

Small Arms Survey. "Darfur Armed Opposition Groups." Human Security Baseline Assessment for Sudan and South Sudan. October 8, 2012. Retrieved May 15, 2019, from http://www.smallarmssurveysudan.org/facts-figures/sudan/darfur/darfurs-armed-groups/darfurs-armed-opposition-groups.html.

Smith, Chris M., and Andrew V. Papachristos. "Trust Thy Crooked Neighbor: Multiplexity in Chicago Organized Crime Networks." *American Sociological Review* 81, no. 4 (2016): 644–67.

Staniland, Paul. *Networks of rebellion.* Ithaca, NY: Cornell University Press, 2014.

Stewart, Megan A. "Civil War as State-Making: Strategic Governance in Civil War." *International Organization* 72, no. 1 (2018): 205–26.

Sundberg, Ralph, Kristine Eck, and Joakim Kreutz. "Introducing the UCDP Non-State Conflict Dataset." *Journal of Peace Research* 49, no. 2 (2012): 351–62.

The Guardian. "Who Is Foday Sankoh?" 2000. https://www.theguardian.com/world/2000/may/17/sierraleone.

Thomas, Jakana. "Rewarding Bad Behavior: How Governments Respond to Terrorism in Civil War." *American Journal of Political Science* 58, no. 4 (2014): 804–18.

Tichy, Noel M., Michael L. Tushman, and Charles Fombrun. "Social Network Analysis for Organizations." *Academy of Management Review* 4, no. 4 (1979): 507–19.

Tokdemir, Efe, and Seden Akcinaroglu. "Reputation of Terror Groups Dataset: Measuring Popularity of Terror Groups." *Journal of Peace Research* 53, no. 2 (2016): 268–77.

Uzzi, Brian. "The Sources and Consequences of Embeddedness for the Economic Performance of Organizations: The Network Effect." *American Sociological Review* 61, no. 4 (1996): 674–698.

Vargas, Robert. "Criminal Group Embeddedness and the Adverse Effects of Arresting a Gang's Leader: A Comparative Case Study." *Criminology* 52, no. 2 (2014): 143–68.

Wagstaff, William A., and Danielle F. Jung. "Competing for Constituents: Trends in Terrorist Service Provision." *Terrorism and Political Violence* 32, no. 2 (2020): 293–324.

Walsh, James I., and James A. Piazza. "Why Respecting Physical Integrity Rights Reduces Terrorism." *Comparative Political Studies* 43, no. 5 (2010): 551–77.

Walter, Barbara F. "Does Conflict Beget Conflict? Explaining Recurring Civil War." *Journal of Peace Research* 41, no. 3 (2004): 371–88.

Walter, Barbara F. "The Extremist's Advantage in Civil Wars." *International Security* 42, no. 2 (2017): 7–39.

Wasserman, Stanley, and Katherine Faust. *Social Network Analysis.* Cambridge, UK: Cambridge University Press, 1994.

Weinstein, Jeremy M. *Inside Rebellion: The Politics of Insurgent Violence.* Cambridge, UK: Cambridge University Press, 2006.

Wood, Elisabeth Jean. *Insurgent Collective Action and Civil War in El Salvador.* Cambridge, UK: Cambridge University Press, 2003.

Wood, Elisabeth Jean. "The Social Processes of Civil War: The Wartime Transformation of Social Networks." *Annual Review of Political Science* 11 (2008): 539–61.

Wood, Reed M., and Jacob D. Kathman. "Competing for the Crown: Inter-Rebel Competition and Civilian Targeting in Civil War." *Political Research Quarterly* 68, no. 1 (2015): 167–79.

Wrong, Dennis H. "The Oversocialized Conception of Man in Modern Sociology." *American Sociological Review* (1961): 183–93.

Zech, Steven T., and Michael Gabbay. "Social Network Analysis in the Study of Terrorism and Insurgency: From Organization to Politics." *International Studies Review* 18, no. 2 (2016): 214–43.

Zenn, Jacob. "Boko Haram's Evolving Tactics and Alliances in Nigeria." *CTC Sentinel* 6, no. 6 (2013): 10–16.

3

Describing the Big, Allied, and Dangerous II Insurgency Data and Other Data Sources

3.1. Introduction

This chapter describes the data used to test the hypotheses through the book about why insurgent groups attack civilians. We introduce the Big, Allied, and Dangerous II (BAAD2) Insurgency data set by first discussing the historical antecedents to the data (BAAD1) and then providing an overview of the groups in the data—140 organizations observed for up to 15 years. The chapter then describes each of the variables used in the quantitative tests. The data are innovative for including information on relationships among insurgent groups (alliances and rivalry); government efforts to suppress specific groups (carrots and sticks); and other group-level variables, including ideology, social service provision, and leadership type.

3.2. Antecedents and the BAAD 2 Insurgency Data

The predecessor of the BAAD2 Insurgency data set did not focus on insurgent organizations but, instead, on terrorist groups. Indeed, the very motivation for creating BAAD1 was an interest in terrorism and terrorist organizations. After the attack against the United States on September 11, 2001, two of the authors (Asal and Rethemeyer) decided to work together to bring their separate knowledge of general political violence and social network analysis to the study of terrorist organizational behavior. When they first started searching for data to analyze the behavior of terrorist organizations, they were very surprised to learn that although a great deal of data on terrorist attacks were available, covering both international (Enders and Sandler 20011) and domestic and international [LaFree and Dugan 2007; Memorial Institute for the Prevention of Terrorism (MIPT) 2006] incidents, there were no data specifically focused on organizational features such as size

Insurgent Terrorism. Victor Asal, Brian J. Phillips, and R. Karl Rethemeyer, Oxford University Press. © University of Maryland National Consortium for the Study of Terrorism and Responses to Terrorism (START) 2022.
DOI: 10.1093/oso/9780197607015.003.0003

or ideology for terrorist organizations globally—certainly none that were in a format that could be easily analyzed statistically.

After searching for an extended period of time, Asal and Rethemeyer came across MIPT's Terrorism Knowledge Base (TKB; MIPT 2006), which as far as they were able to discern was one of the only extant efforts to collect information on such organizations. MIPT's data set was not organized into a case-by-variables format that could be directly analyzed, but it had webpages for each organization with narratives as well as information about key variables related to each organization, such as size and network connections. The data were also focused on terrorist organizations and not on violent non-state actors in general. MIPT's website defined terrorism as "violence, or the threat of violence, calculated to create an atmosphere of fear and alarm. These acts are designed to coerce others into actions they would not otherwise undertake, or refrain from actions they desired to take." Using MIPT as a source, Asal and Rethemeyer used its definition of terrorism and converted the information from the narratives and other material into a data set that could be analyze quantitatively, after correcting and cleaning the data. Given that this work commenced in 2006, Asal and Rethemeyer used the time period from 1998 to 2005 as the coding window, identifying 499 organizations that had been involved in at least one event tracked in the TKB (though many incidents listed the perpetrator as "unknown"). Of those organizations, only 400 had an MIPT webpage with organizational information, and 5 of these organizations were covers for other organizations listed, which resulted in a data set of 395 terrorist organizations active from 1998 to 2005 for at least 1 year. These data allowed Asal and Rethemeyer to analyze the behavior of terrorist organizations as impacted by their size, ideology, state sponsorship, and network connections, along with several other variables. The BAAD1 data set was a productive effort that resulted in 11 articles on topics ranging from why some terrorist organizations are very lethal (Asal and Rethemeyer 2008a) to why some organizations do not kill (Asal and Rethemeyer 2008b) to why some organizations deploy women (Dalton and Asal 2011) to why some terrorist organizations use or pursue chemical, biological, radiological, or nuclear weapons (Asal, Ackerman, and Rethemeyer 2012), among other questions.

When Asal and Rethemeyer were funded by the Center for the Study of Terrorist and Responses to Terrorism at the University of Maryland (a U.S. Department of Homeland Security Center of Excellence) to extend the BAAD1 data to a yearly data set, they had to make some key decisions. One was whether to focus only on entities defined as terrorist organizations or whether to also code for insurgent organizations. In the end, Asal and Rethemeyer decided to code insurgent organizations as well as terrorist organizations (some organizations fit into both categories). There were fewer insurgent organizations than terrorist organizations to code, so analysis could begin much sooner. The key distinction of the BAAD2 data effort at this point was not the coding of insurgent organizations or terrorist organizations—various efforts have engaged with this coding (Cunningham, Gleditsch, and Salehyan 2013; Horowitz and Potter 2014; Sundberg, Eck, and Kreutz 2012)—but, rather, the fact that data were coded on a yearly basis, which few had done before. As coding began on insurgent organizations, it became increasingly apparent that numerous organizations also used terrorism, if terrorism is defined as killing civilians. This realization led Asal and Rethemeyer to think about the importance of insurgent organizations' decision to engage in terrorism, which led to the writing of this book.

3.2.1. Identifying Insurgent Organizations to Include

First, however, Asal and Rethemeyer needed to decide how to identify insurgent organizations. To do so, they turned to one of the best collections of data on insurgent organizations: the Uppsala Conflict Data Program (UCDP). UCDP collects data on violent non-state actors (Pettersson and Wallensteen 2015; Sundberg et al. 2012) but not yearly data on group attributes. Based on the identification of organizations in the UCDP that had been involved in at least 25 battle-related fatalities in any year from 1998 to 2012, Asal and Rethemeyer created a list of potential organizations to code.

Asal and Rethemeyer then reviewed the list of organizations, removing several for various reasons. If the organization was listed as a generic insurgent without a name, then it was removed. Additionally, the coding team could only gather data on organizations that could be clearly identify. For example, when it comes to the conflict between Israel and various Palestinian groups, it is often the case that the UCDP data set identifies the group as the Palestine Liberation Organization or Hamas or the Rejectionist Front. There are, however, occasions when the group is identified only as "Palestinian insurgents." In those cases, the team could not code "Palestinian insurgents" as a separate group. The same is true when the UCDP data set generically refers to Syrian rebels. Occasionally the UCDP battle death data set includes a government as a conflict party, even though the data set focuses on insurgent organizations. When that was the case, the coding team excluded the government entity. An example is the conflict between India and Pakistan in relation to Kashmir.

Once Asal and Rethemeyer removed organizations that were either generic or governmental, they had a list of 158 organizations that were potentially codeable as insurgent organizations. When the coding team began its work, they identified 18 organizations for which there was simply not enough data to code the entity in any meaningful way. An example of such an organization is the Impartial Defense and Security Forces—Ivory Coast. In the end, the coding team was able to generate data on 140 organizations in the data set from 1998 to 2012. These are the organizations that we analyze in this book (Pettersson and Wallensteen 2015; Sundberg et al. 2012).

We note several other aspects about the coding of the organizations. Not all organizations were active for the entire period of time in the data set, and so they were not coded for every year. Some of the organizations were active before 1998, but because our start year for coding was 1998, the coding team gathered information on them from that year on. Asal and Rethemeyer also identified several groups that were active before they were listed in UCDP. In these cases, Asal and Rethemeyer started coding the organizations from the year in which evidence was first found of their activity during the 1998 to 2012 window, and not just based on what UCDP indicated.

Asal and Rethemeyer coded organizations even if they were not committing violence, as long as they were still extant, unless one of four things happened. First, an organization could disband before 2012. Some organizations stated that they were going to disband or made peace agreements with the government but then continued to be active or resumed activity. In these cases, the coding team continued to collect data about them. Second, Asal and Rethemeyer stopped coding organizations in the data set if they made a deal with the government and negotiated a power-sharing or peace agreement and joined the government and thus stopped being an insurgent organization. Organizations that did not make a deal with the government but did transition to nonviolence and remained so for the time period in question were removed from the data after that year. If organizations made a deal with the government and/or embraced nonviolence but returned to committing violence before the end of 2012, Asal and Rethemeyer coded them for the entire time period, including when they were not engaged in violence. Finally, Asal and Rethemeyer stopped coding organizations when they could find no information on their activity for subsequent years, because this suggests dormancy or group death. Overall, the BAAD2 insurgency data advance our ability to analyze questions related to violent non-state actors by providing yearly data, by expanding the time period covered by the data set by a significant number of years, and by adding a large number of new variables. The greatest innovation, compared to other data sets, is the many group-attribute variables that change over time.

3.3. Overview of the Data

The number of group-years in the data set is 1,386, although most reported models have slightly fewer observations due to missing data on country variables. In the first year, there are 76 organizations in the data set, and over time this number increases so that by 2003 there are 91 organizations and in the last year (2012) there are 107 organizations in the data set. Much of this increase over time is associated with "War on Terror"-related insurgencies in countries such as Afghanistan and Iraq and subsequent related conflicts in countries such as Nigeria, Pakistan, and Syria. We note that our data set includes a substantial number of Islamist organizations due to the years we are examining. If we had data from before 1998, the percentage of Islamist groups would be lower. As a result, some caution should be used if attempting to apply our findings to earlier years.

The 140 groups in the data are diverse in many ways. There is substantial regional variation. As Figure 3.1 indicates, sub-Saharan Africa and Asia have the most insurgent groups—52 and 46, respectively. The Middle East and North Africa also contain a substantial number of organizations—35. Europe and the Americas contain far fewer insurgent organizations—5 and 4, respectively. This is consistent with regional data on civil conflicts (Pettersson et al. 2019). The groups are also quite different with regard to how long they last, or at least how long they are observed in the data. Figure 3.2 shows that approximately 40% of the groups (57) are in the sample for all 15 years. Only 36 groups are in the data for less than 5 years, suggesting that most groups are observed for a substantial number of years, allowing for interesting variation over time.

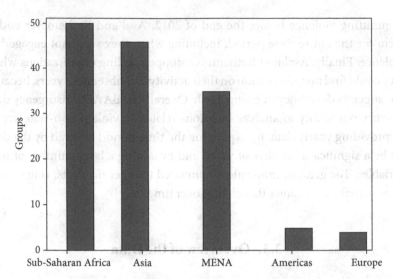

Figure 3.1 Number of insurgent groups by region. MENA, Middle East and North Africa.

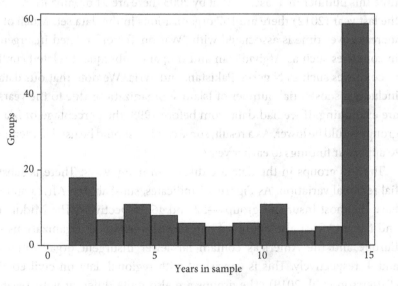

Figure 3.2 Number of groups and years in the data sample.

3.3.1. Variables

Table 3.1 provides summary statistics for all the variables we are using in the analysis. The dependent variable, *Terrorism*, is a dichotomous variable coded 1 if, according to the Global Terrorism Database (GTD), the group attacked civilians in a given year. If the organization only attacked military or police targets, according to the GTD, but did not attack civilians, it was

Table 3.1 Summary Statistics of Variables

Variable	N	Mean	Standard Deviation	Minimum	Maximum
Terrorism	1,240	0.405	0.491	0	1
Terrorism (including military and police targets)	1,240	0.454	0.498	0	1
Lethal terrorism (including military and police targets)	1,240	0.392	0.488	0	1
Lethal terrorism	1,240	0.336	0.473	0	1
Terrorism (count)	1,240	6.935	27.415	0	419
Carrot	1,240	0.045	0.208	0	1
Stick	1,240	0.453	0.498	0	1
Alliances	1,240	1.845	3.151	0	17.5
Rivalry	1,240	0.337	0.658	0	4
Social services	1,240	0.099	0.299	0	1
Ethnic motivation	1,240	0.551	0.498	0	1
Crime	1,240	0.548	1.012	0	5
Mixed carrot and stick	1,240	0.240	0.427	0	1
Territory	1,240	0.245	0.430	0	1
Religious motivation	1,240	0.348	0.477	0	1
Leftist motivation	1,240	0.218	0.413	0	1
Size	1,240	2.689	0.698	1	4
Single leader	1,240	0.055	0.228	0	1
Terrorism history	1,240	28.415	86.177	0	1152
Conflict battle deaths	1,240	209.077	830.309	0	14716
State terror	1,240	4.042	0.884	1	5
Democracy	1,240	5.149	2.837	0.25	10
GDP per capita	1,240	6.770	1.102	4.151	10.791
Population (log)	1,239	17.749	1.629	13.475	20.936

GDP, gross domestic product.

coded 0. Interestingly, only a minority of the group-years, approximately 40%, are coded 1 for terrorism. This type of variation is one of the reasons we use a dichotomous dependent variable instead of a count dependent variable. The key variation seems to be between whether a group uses terror or not instead of how many acts a group launches in a year. Relatedly, we are more confident that the GTD correctly identifies the groups using terrorism than in the accuracy of every recorded attack. Other scholars have used binary dependent variables in similar studies (e.g., Polo 2020; Stanton 2013). We also test alternate operationalizations of terrorism, discussed later.

There is substantial variation in the dependent variable across groups and across years. Figure 3.3 shows how many insurgent groups never used terrorism, used terrorism in some years but not all, or in every year. Only 38 groups (27%) never use terrorism. Examples of these groups are the Myanmar National Democratic Alliance Army and the West Side Boys (Sierra Leone). Most groups (87, or 62%) use terrorism in at least 1 year but not in all years. These groups include Hizballah, the Moro National Liberation Front (Philippines), and the South Sudan Liberation Army, to name a few. Only 15 (11%) insurgent groups use terrorism every year. Examples include Abu Sayyaf Group (ASG; Philippines), the Armed Islamic Group (Algeria), and Tehrik-i-Taliban Pakistan. Figure 3.4 shows a more detailed illustration, with

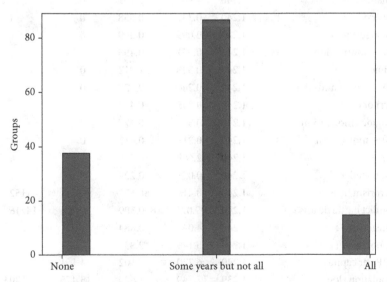

Figure 3.3 Number of groups that used terrorism in none of the years, sometimes, or in all years.

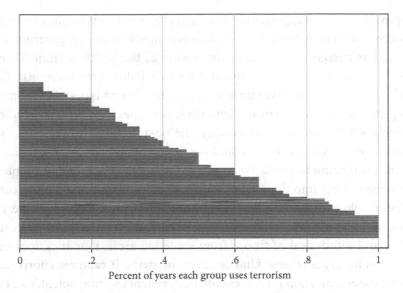

Figure 3.4 Number of times each insurgent group is coded as using terrorism.
Note that each bar represents a group. The white space at the top left represents the 38 groups that did not use terrorism any year.

one bar depicting each group. Each bar indicates the percentage of years that the group used terrorism. Figure 3.4 demonstrates that there is significant variation across groups in terms of how many years they have used terrorism.

In addition to our main dependent variable, we also use some alternative measures of terrorism as outcomes of interest. Our primary measure is dichotomous and is coded 1 only if there are attacks against civilian (not military or police) targets. However, we also report models with the following types of dependent variables: any attack in the GTD, including those against military or police targets; any attack in the GTD resulting in at least one fatality, including those against military or police targets; or any terrorist attack in the GTD resulting in at least one fatality, excluding military or police targets. These are all dichotomous measures. In addition, we include models with *count* versions of all the previous dependent variables and the main dependent variable. The count models use negative binomial regression, as is commonly done in the literature.

Regarding independent variables, the data include a broad range of organizational characteristics. To examine the impact of counterinsurgency and counterterrorism efforts, we code variables that cover varying types of such efforts—particularly conciliatory (carrot) and coercive (stick) approaches. We are aggregating the efforts of all actors (excluding non-state groups) countering the insurgents, such as domestic or foreign governments as well as intergovernmental institutions (e.g., the North Atlantic Treaty Organization or the International Criminal Police Organization). The coding for these efforts was done to aggregate different types of counterinsurgency and counterterrorism into three key types of efforts. We studied specific efforts that were focused on the organizations in questions, although certain kinds of efforts can have spillover effects impacting civilians in the areas being targeted. We divide the counterinsurgency and counterterrorism efforts into three types: carrot, stick, or mixed. We capture conciliatory efforts directed at the organization through the carrot variable. By this, we mean peace talks, release of prisoners, efforts to talk before peace talks, and withdrawal of troops from contested areas. Our stick measure captures the exact reverse kind of efforts by states. It captures efforts that target the organization for repression, both violent and nonviolent, and the use of military attacks and other efforts to negatively affect the organization. Note that at times, states or other actors will adopt both carrot and stick strategies in the same year and sometimes even at the same time. To capture this behavior, we used the mixed variable. This can also be the result of different organizations applying different strategies at the same time. Stick strategies are most common: Only approximately 4% of the group-years are coded for *Carrot*, whereas 45% of the group-years are coded for *Stick* and 23% are coded for *Mixed*.

To test Hypotheses 3 and 4, we include measures of group alliances and rivalry. Interorganizational alliances can provide groups with resources to carry out terrorism (Asal and Rethemeyer 2008; Bacon 2014), whereas intergroup competition often has deleterious consequences for civilians, as militant groups use terror to signal their prowess or resolve (Bloom 2005; Conrad and Greene 2015; Metelits 2009; Nemeth 2013). For these variables, information was compiled into a dyadic network that included only those insurgent organizations active in that year. We then calculated several measures of network centrality to gauge organizations' intensity of relations with other insurgent entities. Because there is no consensus on the "best" measure of network connectivity in this context, we experimented with several measures. Previous research (Asal and Rethemeyer 2008; Asal et al. 2012; Horowitz and Potter 2014) has demonstrated that different phenomena are better explained by different forms of centrality. Here, more sophisticated measures of network *embeddedness* (Granovetter 1985) provide a better fit.

Alliances were measured through average reciprocal distance (ARD) and calculated using UCINET 6's "multiple centrality" command for each year. ARD is a modified form of closeness centrality; it measures the number of alliance connections but also allows for more distant alliances to count (unlike degree centrality, another measure we calculated and used in our analysis). ARD has attractive properties for disconnected networks (Borgatti, Carley, and Krackhardt 2006; Cunningham, Everton, and Murphy 2016). Groups with alliances at the mean level include the National Liberation Army of Colombia and the National Democratic Front of Bodoland during many years and Hamas in the late 1990s. The group with the highest level of alliance embeddedness is al-Qaeda in the late 2000s.

Rivals are groups in competition against each other. The BAAD codebook specifically states that rivalry is when "organizations compete for the same object or goal as another, or try to equal or outdo another; competitor. They seek to dispute another's preeminence or superiority" (Asal, Rethemeyer, and Weedon 2015, 23). We use a simple count of rivals to measure rivalry because we are more interested in the number of advesaries than assessing how important an enemy of an enemy is. It is unclear that rivals of rivals will have important effects on insurgent groups in the same way that allies having allies can be fortuitous in a number of ways (Horowitz and Potter 2014). Rivalries are somewhat rare, as approximately 75% of the group-years record no rivalries. Regarding particular groups, 30% of the groups have at least one rival in at least 1 year of the data, and some have multiple rivals. Groups with at least two rivals in some years include Hamas, the Moro National Liberation Front (Philippines), and the Azawad National Liberation Movement (Mali). It is possible for two groups to have elements of a positive and negative relationship within 1 year (Bencherif and Campana 2017; Berti 2020). In these cases, such as when rivals Fatah and Hamas also collaborate, both rival and allied relationships are recorded.

To measure Hypothesis 5, we include a variable to indicate if the insurgent organization is providing social services to some civilian population. The variable is binary and is coded as a 1 if there is evidence in a year that the organization in question provides any medical, welfare, education, infrastructure, or protection services. During the time period in question, 31 organizations (22%) provided social services in at least 1 year. The following three organizations, however, provided social services for the entire time period of the data set (1998 to 2012): Hamas, Hizballah, and the Taliban. Other groups that provided social services during multiple years include al

Shabaab, West Side Boys (Sierra Leone), and the Moro Islamic Liberation Front (Philippines).

Ethnic motivation is a dichotomous variable included to capture Hypothesis 6 about groups' relationships with the local community. An organization was coded as ethnic if it claimed to represent an ethnic group and if it advocated for the rights or expansion of the rights or territory of that ethnic group. This variable derives from a broader set of ideology variables discussed later with the control variables *Religious motivation* and *Left motivation*. Groups can be coded for multiple ideologies, and many are. Some groups coded exclusively for *Ethnic motivation* include the Baloch Republican Army (Pakistan), Fatah, the Movement of Democratic Forces of Casamance (Senegal), the Real Irish Republican Army (RIRA), and the Sudan People's Liberation Army. Groups that were coded for ethnic motivation as well as for religious and/or leftist motivations include Hamas, the Justice and Equality Movement (Sudan), Palestinian Islamic Jihad, and the United Liberation Front of Assam (India). We use this broad notion of ethnic motivation, where ethnic motivation could be compounded with other ideologies, because we are interested in the *existence* of ethnic motivation not its purity or singularity as a motivating ideology for an organization. However, as a robustness check, we also look at exclusive ethnic motivations. Almost half (68) of the groups are coded 1 for *Ethnic motivation*.

For Hypothesis 7, we use *Crime*, a count variable indicating the number of types of criminal activity each group was involved in during a given year. An organization was coded 1–5 based on whether it was involved in any of the following activities in a year: drug trafficking, smuggling, robbery, kidnapping, or extortion. We use a count instead of a dichotomous variable as our primary measure because there is a great deal of variation in the degree to which groups are involved in crime, and involvement in multiple types of crime generally suggests more criminal activity. In addition, and crucially for the theoretical idea we are trying to capture (criminal embeddedness in society), the more types of crime a group perpetrates, the more elements of society the group affects through its crime. Slightly less than 30% of the group-years, (408), had an organization that was involved in at least one of these behaviors. In approximately 15% (221) of the group-years, an organization was involved in one criminal activity, whereas in less than 1% of the years (13) an organization was involved in all five. All but 2 of these 13 group-years belong to the ASG, whereas RIRA and al-Qaeda in the Lands of the Islamic Maghreb were able to achieve this for only 1 year each. Regarding groups instead of group-years, more than half of the groups (74) were involved in at least one type of crime in at least 1 year.

In addition to the variables representing hypothesized relationships, we include in models many variables to control for alternate explanations. Regarding the coercive and conciliatory state action variables *Carrot* and *Stick*, we also include a variable indicating years when the group was subject to both approaches. The *Mixed* variable was described previously.

Territorial control has been shown to be important for a number of insurgent behaviors (Asal et al. 2016; Asal and Jadoon 2020; Asal, Rethemeyer, and Schoon 2019; Toft 2010). With respect to terrorism, some work finds that territorial control can be positively associated with civilian targeting (Abrahms and Potter 2015), whereas other work finds a negative relationship (Stewart and Liou 2017; Wood 2014), and still other work finds no relationship (Polo 2020; Polo and Gleditsch 2016). To control for either a positive or a negative relationship, we coded territorial control for each organization based on certain criteria that showed that an organization had real control over an area. A key factor that illustrates control is if the organization controlled movement into or out of the territory. The territory may be controlled by force and the threat of violence, through popular support and voluntary submission, or if the government has granted control of that territory to the organization. The key element is if the organization controls the territory not connected to whatever means it needs to use to do so. To be coded as controlling territory, the organization needs to control not just a city block but a significant area such as a whole city or a region. Control also needs to be for an extended period of time, usually multiple weeks. In our data, there were 342 organizational years (almost 25% of the group-years) in which an organization controlled territory. Approximately half (71) of the groups control territory in at least 1 year.

As noted previously, we coded the ideology/ideologies of each group and include *Ethnic motivation* as a key independent variable. Two other ideologies we use as control variables: *Religious motivation* and *Leftist motivation*. An organization was coded as religious if it views itself as being guided by certain religious principles and the need to incorporate religious policies. An example of such an organization is ASG. Leftist organizations were coded as such if they proclaimed an ideology that pushed for the redistribution of wealth to the poor or if they wanted the nationalization of industry or both. An example of such an organization in the data set is the Communist Party of Nepal–Maoist. *Religious motivation* could be associated with more terrorism

(Asal and Rethemeyer 2008), although few studies have considered how religion might affect insurgent groups in particular. Leftist goals could be negatively related to terrorism because some work finds leftist militant groups
carry out fewer attacks than other types of groups (Piazza 2009). Forty-eight
groups are coded as religious, whereas 24 are coded as leftist.

Group size, in terms of members, is a commonly used proxy for group
strength (e.g., Daxecker and Hess 2013; Hou, Gaibulloev, and Sandler
2020). Given substantial discrepancies in estimates of group membership,
and no precise data for many groups, we measure size with an ordinal variable. *Group size* is coded 1 for groups with fewer than 100 militants, 2 for
groups with between 100 and 999 members, 3 for groups between 1,000 and
9,999 members, and 4 for groups with at least 10,000 members. The most
common category for *Group size* is between 1,000 and 9,999 members, with
76 groups—just over half—in this range. The distribution of group sizes is
shown in Figure 3.5. It is unclear how group size might affect insurgent terrorism. On the one hand, terrorism is often argued to be a weapon of the
weak, and some studies of insurgent terrorism find weakness associated
with civilian targeting (Stanton 2013; Wood 2010). On the other hand, some
studies find that capability, in some contexts, leads to more violence against
civilians (Asal and Rethemeyer 2008a; Wood 2014). Other work finds no relationship between group strength and civilian targeting (Keels and Kinney

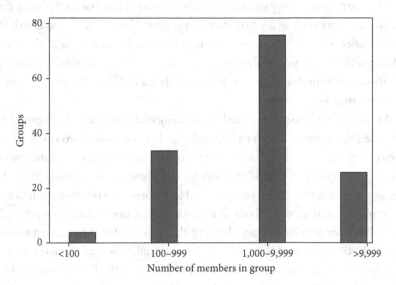

Figure 3.5 Distribution of group size.

2019). However, we think it is reasonable that group strength should have an effect, per popular theoretical arguments, so this variable seems important to include as a control.

To control for the possible impact of leadership type on the targeting of civilians, we use a dichotomous variable *Single leader*, coded 1 for the 13 groups (approximately 10% of the total) each led by a single person. These groups might be especially unlikely to use terrorism because extant research suggests groups with centralized leadership are less likely to do so (Abrahms and Potter 2015). A growing body of literature highlights the importance of leaders to insurgent group dynamics (e.g., Ingram 2016; Lutmar and Terris 2019; Tamm 2016), although this literature has not been sufficiently integrated with the research on insurgent terrorism beyond the Abrahms and Potter study. To take into consideration past group behavior, we include a variable called *terrorism history*, which is a cumulative count of attacks attributed to the group in the GTD without military or police targets. This measure may signal to the state and population that a group is capable and willing to continue this tactic. Beyond group attributes, models include the number of battle deaths each year in the conflict (UCDP 2016) to control for broader conflict dynamics. Higher battle deaths should be associated with higher levels of civilian targeting (Fortna, Lotito, and Rubin 2018).

Models also include control variables for several state-level factors, all of which were extracted from the Quality of Government data set (Teorell et al. 2020). *Political terror* is Gibney, Cornett, and Wood's (2010) measure of state-sanctioned killing, disappearances, and similar repression. We use the Amnesty International version of the variable, but if the State Department version is used, results are similar. This variable could be associated with insurgent terrorism if the state and rebels engage in tit-for-tat terror (Bell and Murdie 2018; Danneman and Ritter 2014), more broadly than the repression captured in *Stick*. *Regime type* is included because some work finds that democracy is negatively related to civilian targeting by insurgents (Jo and Simmons 2016), whereas other work finds the opposite (Eck and Hultman 2007; Hultman 2012; Stanton 2013). We use the Freedom House imputed measure based on Polity data to minimize missing observations, which is substantial if either Freedom House or Polity is used on its own. (Robustness checks use alternate regime type measures and return similar results.) Models also include the natural log of gross domestic product per capita. The source is the World Bank World Development Indicators. Some work finds that insurgents in richer states tend to attack civilians more often (Gleditsch and Polo 2016; Keels and Kinney 2019; Stanton 2013), perhaps because insurgents in wealthy states are less capable of directly attacking the relatively strong state military wealth affords. Finally, models include a logged version of national population to take into consideration that some countries have much larger populations than others. Some research suggests that larger countries are more likely to experience insurgent violence and/or terrorism than others (Chenoweth 2010).

3.4. Conclusion

This chapter described the yearly data on 140 insurgent organizations from 1998 to 2012. The newer edition of the BAAD data, BAAD2 Insurgency data, is innovative for focusing on insurgent organizations as opposed to only terrorist organizations, as the BAAD1 data did. Among data sets on insurgent organizations, BAAD2 stands out for its group-level variables and their yearly variation. The data provides information on intergroup relations (alliances and rivalries), tactics used specifically against the group by the government (carrots and sticks), and broader factors indicating relations with the community (social services, ethnic motivation, and involvement in crime) that together help us test our theoretical argument. The data also includes important group-level control variables such as territorial control, ideology, and leadership type. These controls are important innovations because many studies of insurgent terrorism have not included, for example, measures of group ideology or leadership type (Fortna, Lotito, and Rubin 2018; Stanton 2013; Wood 2014). Chapter 4 describes models using these variables to test the hypotheses presented in Chapter 2.

References

Abrahms, Max, and Philip B. K. Potter. "Explaining Terrorism: Leadership Deficits and Militant Group Tactics." *International Organization* 69, no. 2 (2015): 311–42.

Asal, Victor, and Amira Jadoon. "When Women Fight: Unemployment, Territorial Control and the Prevalence of Female Combatants in Insurgent Organizations." *Dynamics of Asymmetric Conflict* 13, no. 3 (2020): 258–81.

Asal, Victor H., Gary A. Ackerman, and R. Karl Rethemeyer. "Connections Can Be Toxic: Terrorist Organizational Factors and the Pursuit of CBRN Weapons." *Studies in Conflict & Terrorism* 35, no. 3 (2012): 229–54.

Asal, Victor H., Hyun H. Park, R. Karl Rethemeyer, and Gary Ackerman. "With Friends Like These . . . Why Terrorist Organizations Ally." *International Public Management Journal* 19, no. 1 (2016): 1–30.

Asal, Victor H., and R. Karl Rethemeyer. "The Nature of the Beast: Terrorist Organizational Characteristics and Organizational Lethality." *Journal of Politics* 70, no. 2 (2008a): 437–49.

Asal, Victor H., and R. Karl Rethemeyer. "Dilettantes, Ideologues, and the Weak: Terrorists Who Don't Kill." *Conflict Management and Peace Science* 25, no. 3 (2008b): 244–63.

Asal, Victor H., R. Karl Rethemeyer, and Eric W. Schoon. "Crime, Conflict, and the Legitimacy Trade-Off: Explaining Variation in Insurgents' Participation in Crime." *Journal of Politics* 81, no. 2 (2019): 399–410.

Asal, Victor H., R. Karl Rethemeyer, and Suzanne Weedon. *Big Allied and Dangerous (BAAD) Codebook.* Albany, NY: Project on Violent Conflict, 2015.

Bacon, Tricia. "Alliance Hubs: Focal Points in the International Terrorist Landscape." *Perspectives on Terrorism* 8, no. 4 (2014): 4–26.

Bell, Sam R., and Amanda Murdie. "The Apparatus for Violence: Repression, Violent Protest, and Civil War in a Cross-National Framework." *Conflict Management and Peace Science* 35, no. 4 (2018): 336–54.

Bencherif, Adib, and Aurélie Campana. "Alliances of Convenience: Assessing the Dynamics of the Malian Insurgency." *Mediterranean Politics* 22, no. 1 (2017): 115–34.

Berti, Benedetta. "From Cooperation to Competition: Localization, Militarization and Rebel Co-Governance Arrangements in Syria." *Studies in Conflict & Terrorism* (2020): 1–19.

Bloom, Mia. *Dying to Kill: The Allure of Suicide Terror.* Columbia University Press, 2005.

Borgatti, Stephen P., Carley, Kathleen M., and Krackhardt, David. "On the Robustness of Centrality Measures Under Conditions of Imperfect Data." *Social Networks* 28, no. 2 (2006): 124–36.

Chenoweth, Erica. "Democratic Competition and Terrorist Activity." *Journal of Politics* 72, no. 1 (2010): 16–30.

Cunningham, David E., Kristian Skrede Gleditsch, and Idean Salehyan. "Non-State Actors in Civil Wars: A New Dataset." *Conflict Management and Peace Science* 30, no. 5 (2013): 516–31.

Cunningham, David, Everton, Sean, Murphy, Philip. *Understanding Dark Networks: A Strategic Framework for the Use of Social Network Analysis.* London, UK: Rowman & Littlefield, 2016.

Dalton, Angela, and Victor H. Asal. "Is It Ideology or Desperation: Why Do Organizations Deploy Women in Violent Terrorist Attacks?" *Studies in Conflict & Terrorism* 34, no. 10 (2011): 802–19.

Danneman, Nathan, and Emily Hencken Ritter. "Contagious Rebellion and Preemptive Repression." *Journal of Conflict Resolution* 58, no. 2 (2014): 254–79.

Daxecker, Ursula E., and Michael L. Hess. "Repression Hurts: Coercive Government Responses and the Demise of Terrorist Campaigns." *British Journal of Political Science* 43 (2013): 559–77.

Eck, Kristine, and Lisa Hultman. "One-Sided Violence Against Civilians in War: Insights from New Fatality Data." *Journal of Peace Research* 44, no. 2 (2007): 233–46.

Fortna, Virginia Page, Nicholas J. Lotito, and Michael A. Rubin. "Don't Bite the Hand That Feeds: Rebel Funding Sources and the Use of Terrorism in Civil Wars." *International Studies Quarterly* 62, no. 4 (2018): 782–94.

Gibney, Mark, Linda Cornett, and Reed Wood. "Political Terror Scale 1976–2009." 2010. Retrieved from http://www.politicalterrorscale.org.

Gleditsch, Kristian Skrede, and Sara M. T. Polo. "Ethnic Inclusion, Democracy, and Terrorism." *Public Choice* 169, nos. 3–4 (2016): 207–29.

Granovetter, Mark. "Economic Action and Social Structure: The Problem of Embeddedness." *American Journal of Sociology* 91, no. 3 (1985).

Horowitz, Michael C., and Philip B. K. Potter. "Allying to Kill: Terrorist Intergroup Cooperation and the Consequences for Lethality." *Journal of Conflict Resolution* 58, no. 2 (2014): 199–225.

Hou, Dongfang, Khusrav Gaibulloev, and Todd Sandler. "Introducing Extended Data on Terrorist Groups (EDTG), 1970 to 2016." *Journal of Conflict Resolution* 64, no. 1 (2020): 199–225.

Hultman, Lisa. "Attacks on Civilians in Civil War: Targeting the Achilles Heel of Democratic Governments." *International Interactions* 38, no. 2 (2012): 164–81.

Ingram, Haroro J. *The Charismatic Leadership Phenomenon in Radical and Militant Islamism*. New York, NY: Routledge, 2016.

Jo, Hyeran, and Beth A. Simmons. "Can the International Criminal Court Deter Atrocity?" *International Organization* 70, no. 3 (2016): 443–75.

Keels, Eric, and Justin Kinney. "'Any Press Is Good Press?' Rebel Political Wings, Media Freedom, and Terrorism in Civil Wars." *International Interactions* 45, no. 1 (2019): 144–69.

LaFree, Gary, and Laura Dugan. "Introducing the Global Terrorism Database." *Terrorism and Political Violence* 19, no. 2 (2007): 181–204.

Lutmar, Carmela, and Lesley G. Terris. "Introducing a New Dataset on Leadership Change in Rebel Groups, 1946–2010." *Journal of Peace Research* 56, no. 2 (2019): 306–15.

Memorial Institute for the Prevention of Terrorism. "MIPT Terrorism Knowledge Base." 2006. Retrieved June 15, 2006 from http://www.tkb.org/Home.jsp.

Metelits, Claire. *Inside Insurgency*. New York University Press, 2009.

Nemeth, Stephen. "The Effect of Competition on Terrorist Group Operations." *Journal of Conflict Resolution* 58, no. 2 (2014): 336–62.

Pettersson, Therése, Stina Högbladh, and Magnus Öberg. "Organized Violence, 1989–2018 and Peace Agreements." *Journal of Peace Research* 56, no. 4 (2019): 589–603.

Pettersson, Therese, and Peter Wallensteen. "Armed Conflicts, 1946–2014." *Journal of Peace Research* 52, no. 4 (2015): 536–50.

Piazza, James A. "Is Islamist Terrorism More Dangerous? An Empirical Study of Group Ideology, Organization, and Goal Structure." *Terrorism and Political Violence* 21, no. 1 (2009): 62–88.

Polo, Sara M. T. "How Terrorism Spreads: Emulation and the Diffusion of Ethnic and Ethnoreligious Terrorism." *Journal of Conflict Resolution* 64, no. 10 (2020): 1916–42.

Polo, Sara M. T. "The Quality of Terrorist Violence: Explaining the Logic of Terrorist Target Choice." *Journal of Peace Research* 57, no. 2 (2020): 235–50.

Polo, Sara M. T., and Kristian Skrede Gleditsch. "Twisting Arms and Sending Messages: Terrorist Tactics in Civil War." *Journal of Peace Research* 53, no. 6 (2016): 815–29.

Stanton, Jessica A. "Terrorism in the Context of Civil War." *Journal of Politics* 75, no. 4 (2013): 1009–22.

Stewart, Megan A., and Yu-Ming Liou. "Do Good Borders Make Good Rebels? Territorial Control and Civilian Casualties." *Journal of Politics* 79, no. 1 (2017): 284–301.

Sundberg, Ralph, Kristine Eck, and Joakim Kreutz. "Introducing the UCDP Non-State Conflict Dataset." *Journal of Peace Research* 49, no. 2 (2012): 351–62.

Tamm, Henning. "Rebel Leaders, Internal Rivals, and External Resources: How State Sponsors Affect Insurgent Cohesion." *International Studies Quarterly* 60, no. 4 (2016): 599–610.

Teorell, Jan, Stefan Dahlberg, Sören Holmberg, Bo Rothstein, and Natalia Alvarado. *The Quality of Government Standard Dataset, Version Jan20.* 2020. Gothenburg, Sweden: The Quality of Government Institute, University of Gothenburg.

Toft, Monica Duffy. *The Geography of Ethnic Violence: Identity, Interests, and the Indivisibility of Territory.* Princeton, NJ: Princeton University Press, 2010.

Wood, Reed M. "Opportunities to Kill or Incentives for Restraint? Rebel Capabilities, the Origins of Support, and Civilian Victimization in Civil War." *Conflict Management and Peace Science* 31, no. 5 (2014): 461–80.

Wood, Reed M. "Rebel Capability and Strategic Violence Against Civilians." *Journal of Peace Research* 47, no. 5 (2010): 601–14.

4

Testing Primary Hypotheses

4.1. Introduction

Drawing on the Big, Allied, and Dangerous II (BAAD2) data and variables described in Chapter 3, this chapter describes the model used to test hypotheses presented in Chapter 2. We begin by looking at simple correlations between the dependent variable and the independent variables, using contingency tables. We then conduct a battery of statistical analyses and generally find support for our hypotheses. Measures of embeddedness are consistently associated with insurgent terrorism. "Carrot" conciliatory practices are related to less civilian targeting by insurgent groups, whereas "stick" coercive tactics, alliances, rivalries, social service provision, ethnic motivation, and crime are all associated with more civilian targeting. These results are mostly robust regardless of how "terrorism" is measured. We do not find support for alternative explanations of civilian targeting by insurgent organizations, such as insurgent group weakness or operating in a democratic country.

4.2. Preliminary Analysis: Contingency Tables

Before using multivariate analysis to control for multiple possible confounding factors and other aspects of the data, we show simple contingency tables, also called cross-tabulation or cross-tabs, to determine if relationships seem to exist at the most basic level. Contingency tables can be helpful for exploratory analysis, and they can be used as a robustness check of relationships—that is, to determine if any relationship exists at all before turning to more sophisticated analyses. These tables are frequently used in quantitative analyses of social science phenomena, including organized violence (e.g., Ishiyama et al. 2018; Savun and Gineste 2019).

Insurgent Terrorism. Victor Asal, Brian J. Phillips, and R. Karl Rethemeyer, Oxford University Press. © University of Maryland National Consortium for the Study of Terrorism and Responses to Terrorism (START) 2022. DOI: 10.1093/oso/9780197607015.003.0004

Table 4.1 Government Conciliation and Insurgent Terrorism in 140 Groups, 1998–2012

	No Terrorism	Terrorism
No Concessions	697	495
Concessions	46	8

Note: Variables used are $Carrot_{t-1}$ and Terrorism. $\chi^2 = 15.313$, Pr = .000, $n = 1{,}246$.

Table 4.2 Government Coercion and Insurgent Terrorism in 140 Groups, 1998–2012

	No Terrorism	Terrorism
No Coercion	487	188
Coercion	256	315

Note: Variables used are $Stick_{t-1}$ and Terrorism. $\chi^2 = 95.871$, Pr =.000, $n = 1{,}246$.

Table 4.1 shows the overlap between *Carrot* and *Terrorism*.[1] Of the group-years in which the state had offered conciliatory measures toward the group the previous year (54), only 8 coincided with group use terrorism. There are far more years (46) in which there had been conciliation followed by *no* terrorism the next year. This is consistent with the notion that government concession might reduce terrorism by insurgents. Table 4.2 shows *Stick* and *Terrorism*. Of the 571 group-years in which an organization had been subject to a government coercive action the previous year, most of those group-years (315) experienced subsequent terrorism by the group. Fewer group-years (256) had government coercion followed by no terrorism. Stated another way, groups that were subject to government coercion in one year usually used terrorism the next year. This is suggestive of a positive relationship between state coercion and subsequent terrorism.

The *Alliances* measure is a count, so it is not as straightforward to show in a cross-tab table. However, if we dichotomize *Alliance* where it is coded "1" if the group-year has greater than the mean value of *Alliance* connections (1.703) and "0" otherwise, such a table is possible. Table 4.3 shows that the

---

[1] Technically the table shows the overlap involving $Carrot_{t-1}$, the *Carrot* value for the previous year. However, we avoid the subscripted language in the text for ease of reading.

Table 4.3 Insurgent Alliances and Insurgent Terrorism in 140 Groups, 1998–2012.

	No Terrorism	Terrorism
No Alliance	624	290
Alliance	221	251

Note: Variables used are a dichotomized version of *Alliances$_{t-1}$* and *Terrorism*. $\chi^2 = 60.176$, Pr $= .000$, $n = 1,386$.

Table 4.4 Insurgent Rivalries and Insurgent Terrorism in 140 Groups, 1998–2012

	No Terrorism	Terrorism
No Rivals	616	325
Rivals	127	178

Note: Variables used are a dichotomized version of *Rivals$_{t-1}$* and *Terrorism*. $\chi^2 = 54.305$, Pr $= .000$, $n = 1,246$.

472 group-years with greater than the mean value of insurgent alliances are slightly more likely to coincide with the use terrorism the following year (251) than not (221). This is consistent with the hypothesis that insurgent alliance connections are associated with insurgent terrorism. Table 4.4 shows a similar exercise with *Rivals*, but with the variable dichotomized based on whether the group-year had at least one rival, or not. Of the 305 group-years in which an insurgent group had at least one insurgent rival the previous year, more of the group-years saw terrorism by the group (178) than not (127). Rivalry, like alliances, seems to be associated with subsequent insurgent terrorism.

Regarding social service provision, Table 4.5 indicates that of the 120 group-years in which a group provided such services, most of those groups (78) used terrorism the following year. This is consistent with expectations. Table 4.6 looks at the overlap between *Ethnic motivation* and *Terrorism*. Of the 753 group-years for which a group is coded as having an ethnonationalist ideology, the groups only use terrorism in a minority of these observations (312). The combination of ethnic motivation and *no* terrorism occurs more frequently (441 group-years). This goes against expectations. However,

Table 4.5 Social Service Provision and Insurgent Terrorism in 140 Groups, 1998–2012

	No Terrorism	Terrorism
No Social Services	701	425
Social Services	42	78

Note: Variables used are *Social services$_{t-1}$* and *Terrorism*. $\chi^2 = 33.466$, Pr = .000, $n = 1{,}246$.

Table 4.6 Ethnic Motivation and Insurgent Terrorism in 140 Groups, 1998–2012

	No Terrorism	Terrorism
No Ethnic Motivation	404	229
Ethnic Motivation	441	312

Note: Variables used are *Ethnic motivation* and *Terrorism*. $\chi^2 = 3.994$, Pr = .046, $n = 1{,}246$.

Table 4.7 Crime Involvement and Insurgent Terrorism in 140 Groups, 1998–2012

	No Terrorism	Terrorism
No Crime	604	277
Crime	139	226

Note: Variables used are a dichotomized version of *Crime$_{t-1}$* and *Terrorism*. $\chi^2 = 99.576$, Pr = .000, $n = 1{,}246$.

this simple cross-tab does not take into additional factors that could affect a possible relationship between these variables.[2] Table 4.7 shows *Crime* and *Terrorism*. It uses a dichotomized, instead of count, version of the crime variable for ease of display. Of the 365 group-years in which a group was engaged in at least one type of crime, in the majority of the subsequent years (226), the groups used terrorism. Only in a minority (139) do we see a group

[2] It is also noteworthy that this relationship is the least statistically significant of those shown in contingency tables. Its *p* value of .046 is quite close to the commonly used threshold of .05, whereas the other *p* values shown in the tables are mostly .000.

involved in crime and then *not* use terrorism. This is consistent with the hypothesis. Overall, most independent variables associated with hypothesized relationships show a bivariate correlation with *Terrorism* in the expected direction. The next section uses multivariate regression to take additional factors into consideration.

4.3. Estimation of Regression Models

Because our dependent variable in most models is a binary measure of insurgent organization use of terrorism in the year in question, our primary models use logistic regression. Later models, with the count dependent variables, use negative binomial regression. To take temporal issues into consideration, models include year dummies. If a time trend (year count) variable is included instead of year fixed effects, results are robust. Because each group is measured repeatedly, we include group random effects to capture otherwise unmeasured group attributes (Beck 2001; Bell and Jones 2015). A Hausman test suggests random effects models are not systematically different from fixed effects models. An additional issue with fixed effects is that they cause time-invariant variables to drop out of the models. In addition to our main model, we show various robustness checks, including a lagged dependent variable and alternate dependent variables including count dependent variables.

4.3.1.   Results

Table 4.8 reports the primary models. We first show some baseline models to demonstrate that results in the full models are not driven by the inclusion of certain variables. Model 1 only includes control variables. Model 2 only includes the variables representing hypothesized relationships. Model 3 is the full model, combining both key independent variables and controls. Interestingly, many control variables, such as *Religious motivation* and *Democracy*, are statistically significant in Model 1 but lose statistical significance when the theorized variables are included. This is remarkable because religious ideology and state democracy are competing explanations for terrorism and particularly for terrorism in civil war (e.g., Hultman 2012; Juergensmeyer 2000; Li 2005). This is discussed more later.

Table 4.8 Primary Models: Logit Models of Terrorism by Insurgent Organizations

	Model 1	Model 2	Model 3
	Only Control Variables	Only Key Variables	Main Model
Carrot		−1.354**	−1.352**
		(0.603)	(0.631)
Stick		0.855***	0.929***
		(0.211)	(0.262)
Alliances		0.195***	0.171***
		(0.049)	(0.051)
Rivalry		0.744**	0.602**
		(0.228)	(0.232)
Social services		1.034**	1.162**
		(0.510)	(0.515)
Ethnic motivation		0.692*	0.929**
		(0.411)	(0.431)
Crime		0.664***	0.508***
		(0.136)	(0.138)
Mixed carrot and stick	−0.173		0.197
	(0.240)		(0.316)
Territory	0.821**		0.499
	(0.302)		(0.322)
Religious	0.898**		0.388
	(0.421)		(0.463)
Leftist	0.417		0.807
	(0.542)		(0.550)
Group size	0.074		0.097
	(0.213)		(0.213)
Single leader	−1.118		−0.670
	(0.826)		(0.793)
Terrorism history	0.002		0.002
	(0.003)		(0.004)
Battle deaths	0.001**		0.001**
	(0.000)		(0.000)
State terror	0.246*		0.017
	(0.145)		(0.151)
Democracy	0.140**		0.054
	(0.068)		(0.067)
Per capita income	0.359*		0.288
	(0.188)		(0.184)

Table 4.8 *Continued*

	Model 1	Model 2	Model 3
	Only Control Variables	Only Key Variables	Main Model
Population (log)	0.259**		0.201*
	(0.120)		(0.117)
Constant	−9.999***	−2.129***	−8.617***
	(2.498)	(0.463)	(2.460)
N	1,240	1,240	1,240

Note: Standard errors in parentheses. Models include group random effects and year fixed effects. All independent variables lagged 1 year. $*p < .10, **p < .05, ***p < .01$.

For variables representing hypothesized relationships, results are robust across Models 2 and 3. The coefficient on *Carrot* is negatively signed and sta-tistically significant, suggesting that if a group is offered conciliation in one year, the next year it is less likely to use terrorism. By contrast, the coefficient on *Stick* is statistically significant and positively signed, suggesting that if a group receives coercive action in one year, it is more likely to use terrorism the following year. *Alliances* is statistically significant and positively signed, suggesting that as groups are further embedded in networks of collaborative relationships, they are more likely to use terrorism. *Rivals* is also statistically significant and negatively signed, suggesting that as groups engage in more rivalry, they are more likely to use terrorism. *Social services* is statistically significant and positively signed, suggesting that groups that engage in ser-vice provision are more likely than groups that do not to subsequently use terrorism. *Ethnic motivation* is statistically significant and positively signed, suggesting that groups claiming to represent a particular ethnic group are more likely to use terrorism. Finally, the coefficient on *Crime* is statistically significant and positively signed, suggesting that groups that engage in crim-inal activity such as drug trafficking are more likely than other groups to sub-sequently use terrorism.

Regarding substantive significance, we use marginal effects to calculate estimated effects in a straightforward way, and these estimations are shown in Figure 4.1. All variables representing hypothesized relationships are as-sociated with substantial changes in the likelihood of a group's subsequent terrorism use. The largest effect is associated with *Carrot*. Government concessions are associated with a 17 percentage-point decrease in the

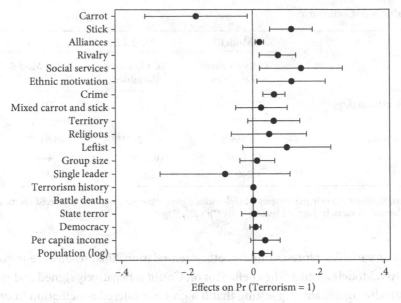

Figure 4.1 Substantive significance of variables in the main model. Average marginal effects are shown with 95% confidence intervals.

probability of subsequent terrorism. To understand this decrease in more practical terms, if all variables are at their means, a group has a 41% probability of experiencing terrorism in a given year. If an insurgent group has received government concessions in the previous year, its probability of using terrorism decreases to 24%, other factors held constant. Other relatively large effects are related to *Social services*, *Stick*, and *Ethnic motivation*, which are associated with 15, 12, and 12 percentage-point increases, respectively, in the probability of subsequent terrorism.

In addition to the variables representing hypothesized relationships, it is noteworthy that few control variables have statistically significant coefficients in the full model. Only *Battle deaths* (the indictor of conflict intensity) and *Country population* are significant at conventional levels (although the latter variable at $p < .10$), and they are positively signed as expected. Groups in civil conflicts that have a higher number of battle deaths are more likely to use terrorism compared to groups in conflicts with lower numbers of battle deaths. Groups in more populous countries are more likely to use terrorism compared to groups in less populous countries.

It is notable that three control variables representing common arguments—*Religious motivation*, *Group size*, and *Democracy*—are all statistically insignificant in the main model, and *Group size* is insignificant in both models. These results are interesting because prominent arguments suggest terrorism is encouraged by religion (Berman and Laitin 2008), group weakness (Wood 2010), or democratic regime types (Hultman 2012; Li 2005; Stanton 2013). It seems that these factors are not related to terrorism if group factors representing embeddedness are taken into consideration. It is also possible that the relationships between these variables and civilian targeting could be more complicated: for instance, the variable may interact with one another or the effect may depend on the particular measurement of the independent variable (Chenoweth 2013; Lee 2013). It is remarkable that *State terror* becomes statistically insignificant in the main model. It is likely that a more direct measure of repression against the group—instead of against other groups or unaffiliated citizens elsewhere in the country—better explains terrorism. The significance of the variable *Stick* lends credence to this idea, suggesting the benefits of using more nuanced group-specific variables instead of country-level measures of repression.

A final interesting finding is that the control variable *GDP per capita* is insignificant. Given the common notion of terrorism as a weapon of the weak, one might expect a greater probability of terrorism in stronger states, and state strength is frequently measured through per capita income (e.g., Fearon and Laitin 2003). However, *GDP per capita* is positively associated with terrorism in the first model, with only control variables ($p < .10$), and statistically insignificant in the full model. Thus, there is no support at all for the notion of terrorism as a weapon of the weak—at least when measured through state per capita wealth—and some marginal support for the idea that wealthier states experience more terrorism. The sample is dominated by relatively poor states, because it is those with insurgent groups present, so any possible relationship between state strength or poverty and terrorism should be considered cautiously. Also note that the measures discussed here are of absolute strength, not relative strength. But the overall lack of support for a link between insurgent weakness and terrorism is striking.

Table 4.9 further checks the robustness of hypothesized relationships using a lagged dependent variable and alternate terrorism measures. The use of lagged dependent variables, particularly with group (fixed) effects, is debated in the literature and potentially problematic (Brandt et al. 2000; Kelle and Kelly 2006). However, because there are debates, we include a lagged dependent variable in Model 4, and results are robust. Model 5 uses the dependent variable *GTD terrorism*, the dichotomous measure of whether a group has carried out any terrorist attacks according to the Global Terrorism Database (GTD), including those against military or police targets. Results

Table 4.9 Robustness Checks: Lagged Dependent Variable and Alternate Dependent Variables

	Model 4	Model 5	Model 6	Model 7
	Lagged DV	Alternate DV: Any GTD Attack (Including Security Targets)	Alternate DV: Any GTD Fatality (Including Security Targets)	Alternate DV: Any Terrorism Fatality
Carrot	−1.361**	−1.472**	−1.134*	−1.422**
	(0.593)	(0.607)	(0.639)	(0.712)
Stick	0.633**	0.838***	0.996***	1.025***
	(0.248)	(0.251)	(0.266)	(0.277)
Alliances	0.114**	0.196***	0.162**	0.130**
	(0.046)	(0.052)	(0.051)	(0.052)
Rivalry	0.343*	0.524**	0.591**	0.574**
	(0.204)	(0.225)	(0.225)	(0.226)
Social services	1.071**	1.188**	0.334	0.196
	(0.429)	(0.527)	(0.500)	(0.490)
Ethnic motivation	0.751**	0.976**	0.650	0.394
	(0.315)	(0.415)	(0.445)	(0.461)
Crime	0.421***	0.622***	0.448***	0.365**
	(0.126)	(0.146)	(0.135)	(0.136)
Mixed carrot and stick	−0.027	−0.031	0.207	0.286
	(0.298)	(0.305)	(0.326)	(0.338)
Territory	0.381	0.495	0.584*	0.852**
	(0.300)	(0.318)	(0.328)	(0.334)
Religious	0.302	0.220	0.380	0.549
	(0.349)	(0.447)	(0.476)	(0.496)
Leftist	0.541	0.742	0.460	0.179
	(0.394)	(0.533)	(0.565)	(0.585)
Group size	0.021	−0.244	−0.251	0.069
	(0.180)	(0.209)	(0.217)	(0.225)
Single leader	−0.600	−0.432	−2.180**	−2.219**
	(0.605)	(0.735)	(0.984)	(1.078)
Terrorism history	0.004	0.000	0.000	0.000
	(0.004)	(0.002)	(0.000)	(0.000)
Battle deaths	0.000**	0.001**	0.001**	0.001**
	(0.000)	(0.000)	(0.000)	(0.000)
State terror	0.008	0.234	0.280*	0.132
	(0.138)	(0.148)	(0.156)	(0.161)

Table 4.9 *Continued*

	Model 4	Model 5	Model 6	Model 7
	Lagged DV	Alternate DV: Any GTD Attack (Including Security Targets)	Alternate DV: Any GTD Fatality (Including Security Targets)	Alternate DV: Any Terrorism Fatality
Democracy	0.066	0.071	0.073	0.095
	(0.054)	(0.067)	(0.068)	(0.069)
Per capita income	0.191	0.355*	0.289	0.196
	(0.142)	(0.182)	(0.192)	(0.198)
Population (log)	0.145*	0.163	0.253**	0.308**
	(0.088)	(0.114)	(0.122)	(0.129)
Lagged DV	1.399***			
	(0.210)			
Constant	−6.684***	−7.814**	−9.175***	−10.174***
	(1.966)	(2.383)	(2.571)	(2.734)
N	1,240	1,240	1,240	1,240

Note: Standard errors in parentheses. Models include group random effects and year fixed effects.
$*p < .10, **p < .05, ***p < .01$.

DV, dependent variable; GTD, Global Terrorism Database.

are consistent with those of the main model, Model 3. Models 6 and 7 use dependent variables based on fatalities instead of only attacks. The dependent variable of Model 6 is *GTD fatalities*, coded 1 if the group has conducted any attacks according to the GTD that killed people. Model 7 uses *Terrorism fatalities*, coded similarly but excluding attacks against military or police targets. Results from Models 6 and 7 are similar to those of other models, with two interesting exceptions. The coefficients on *Social services* and *Ethnic motivation* are statistically insignificant only in these models. This is interesting because both measures are indictors of groups' relationships with the wider community. These types of community relations are associated with an increased likelihood of terrorism (Model 3) but not necessarily *lethal* terrorism (Models 6 and 7). This could indicate some degree of restraint on the part of groups—that they are using violence to maintain control over the population but they do not want to go too far with lethal violence. This is worthy of future research.

There are also two changes regarding control variable results in the models with fatality dependent variables. *Territorial control* is statistically significant and positively signed in Models 6 and 7, suggesting that groups that hold territory are more likely than groups that do not hold territory to use lethal terrorism. This could be because as they work to maintain control of their territory, or expand and increase their territorial control, they kill dissidents or informers who might work with the opponents competing for the territory (e.g., De la Calle 2017; Kalyvas 2006). In addition, *Single leader* is statistically significant and negatively signed in Models 6 and 7, suggesting that groups with a lone leader are less likely to use lethal terrorism. It is possible that these groups are not powerful enough to carry out lethal terrorism, and a more formal leadership structure (e.g., leadership council) is necessary for consistent lethal terrorism.

Table 4.10 shows several additional models, with count instead of dichotomous dependent variables. Model 8 uses a dependent variables that counts the group's terrorist attacks (as before, not including military or police targets). Model 9's dependent variable is a count of any GTD attacks associated with the group – including military or police targets. Model 10 and 11 use counts of terrorist attack *fatalities*, where Model 10 uses the more strict measure excluding attacks on military or police targets, while Model 11's dependent variable is a count of fatalities from any GTD attacks. Results are overall robust with those of the main model (Model 3) and other models, but some interesting differences emerge as well. First, the results for *Carrot* are not as robust when the dependent variable is counts of terrorism—only Model 8 has a statistically significant coefficient, and only at $p < .10$. This suggests that state conciliatory actions are associated with a lower likelihood of subsequent terrorism (Model 3) but not necessarily *fewer* terrorist attacks (e.g., 20 instead of 10). The coefficients on *Ethnic motivation* are statistically insignificant in Models 10 and 11, suggesting that ethnic motivation is associated with an increased likelihood of terrorism, and higher counts of terrorism generally, but not higher counts of *lethal* terrorism. This is consistent with the logic mentioned previously that perhaps groups with ethnic goals want to intimidate their broader ethnic community or a rival ethnic group, but perhaps not often with a high amount of lethal terrorism. There might be diminishing returns to ethnic violence.

Another difference between the main model and the count models is that *Territorial control* is statistically significant in Models 10 and 11. This suggests that although territorial control is not associated with a higher likelihood of terrorism (Model 3) or larger numbers of attacks (Models 8 and 9), it is associated with a higher likelihood of lethal terrorism and additional lethal

Table 4.10 Robustness Checks: Count Dependent Variables

	Model 8	Model 9	Model 10	Model 11
	Count of Terrorist Attacks	Count of Any GTD Attacks (Including Security Targets)	Count of Terrorist Attack Fatalities	Count of Any GTD Attack Fatalities (Including Security Targets)
Carrot	−0.705*	−0.628	−0.476	−0.756
	(0.422)	(0.394)	(0.410)	(0.470)
Stick	0.881***	0.840***	0.876***	0.797***
	(0.156)	(0.143)	(0.159)	(0.177)
Alliances	0.083***	0.079***	0.069**	0.055**
	(0.020)	(0.019)	(0.021)	(0.024)
Rivalry	0.227**	0.280***	0.232**	0.157*
	(0.085)	(0.080)	(0.087)	(0.094)
Social services	0.369**	0.344**	0.470**	0.348**
	(0.156)	(0.154)	(0.167)	(0.177)
Ethnic motivation	0.417**	0.336**	0.144	0.137
	(0.178)	(0.167)	(0.156)	(0.173)
Crime	0.179***	0.180***	0.164***	0.131**
	(0.046)	(0.044)	(0.049)	(0.054)
Mixed carrot and stick	0.488**	0.454**	0.406**	0.394**
	(0.173)	(0.160)	(0.178)	(0.196)
Territory	0.148	0.077	0.255*	0.405**
	(0.135)	(0.128)	(0.140)	(0.150)
Religious	0.118	−0.011	−0.136	0.042
	(0.186)	(0.173)	(0.167)	(0.186)
Leftist	0.433**	0.346*	0.171	0.058
	(0.215)	(0.196)	(0.184)	(0.204)
Group size	0.128	0.023	0.087	0.200**
	(0.092)	(0.086)	(0.090)	(0.100)
Single leader	−0.519	−0.528	−1.489**	−1.430**
	(0.483)	(0.435)	(0.517)	(0.604)
Terrorism history	0.000	0.000	−0.000	−0.000
	(0.000)	(0.000)	(0.000)	(0.000)
Battle deaths	0.000**	0.000**	0.000***	0.000***
	(0.000)	(0.000)	(0.000)	(0.000)
State terror	0.076	0.122*	0.231**	0.214**
	(0.075)	(0.070)	(0.077)	(0.088)

Continued

Table 4.10 *Continued*

	Model 8	Model 9	Model 10	Model 11
	Count of Terrorist Attacks	Count of Any GTD Attacks (Including Security Targets)	Count of Terrorist Attack Fatalities	Count of Any GTD Attack Fatalities (Including Security Targets)
Democracy	0.003	0.003	0.063**	0.052*
	(0.029)	(0.028)	(0.026)	(0.028)
Per capita income	0.141*	0.164**	0.340***	0.362***
	(0.080)	(0.073)	(0.073)	(0.083)
Population (log)	0.088*	0.084**	0.185***	0.186***
	(0.046)	(0.042)	(0.043)	(0.047)
Constant	−5.324***	−5.013***	−9.417***	−9.794***
	(1.153)	(1.045)	(1.037)	(1.174)
N	1,240	1,240	1,240	1,240

Note: Standard errors in parentheses. Models include group random effects and year fixed effects.
*$p < .10$, **$p < .05$, ***$p < .01$.
GTD, Global Terrorism Database.

attacks. Because the literature already finds interesting links between terri-torial control and terrorist violence (Bhavnani, Miodownik, and Choi 2011; Kalyvas 2006; Raleigh 2012; Wood 2014), this is worthy of future research. An additional change is that *Leftist motivation* is statistically significant and positively signed in Models 8 and 9, suggesting groups with leftist goals carry out more terrorist attacks compared to groups with other types of goals. This is interesting, but it is noteworthy that the coefficients for *Leftist motivation* are not statistically significant for any of the fatality measures. Finally, with lethality counts, *Single leader* is statistically significant and negatively signed as it was with the lethality dichotomous measures.

Table 4.11 reports a final set of robustness checks, this time with alternate independent variables. Model 12 includes region control variables, with di-chotomous variables indicating if a group is located in the Americas, Asia, the Middle East and North Africa, or sub-Saharan Africa. In this model, the Middle East and North Africa is the excluded category. To conserve space, the region controls are not shown in the table, but none of the coefficients are statistically significant. Importantly, the results for the hypothesized relationships are not changed by the inclusion of these regional controls.

Models 13 and 14 show alternate measures of regime type. Most models in-clude a measure using imputed Freedom House data based on Polity values,

Table 4.11 Robustness Checks: Alternate Independent Variables

	Model 12	Model 13	Model 14	Model 15	Model 16
	Region Controls Included (Not Shown for Space Reasons)	Alternate Regime Type Measure: Polity2	Alternate Regime Type Measure: Cheibub et al.	Alternate *Ethnic motivation* Measure	Alternate *Crime* Measure
Carrot	−1.366**	−1.172*	−1.532**	−1.306**	−1.410**
	(0.628)	(0.644)	(0.772)	(0.630)	(0.636)
Stick	0.921***	1.041***	1.167***	0.933***	0.876***
	(0.260)	(0.280)	(0.307)	(0.262)	(0.266)
Alliances	0.173***	0.198***	0.226**	0.164**	0.174***
	(0.051)	(0.055)	(0.072)	(0.051)	(0.051)
Rivalry	0.559**	0.704**	0.319	0.633**	0.639**
	(0.234)	(0.248)	(0.262)	(0.231)	(0.235)
Social services	1.184**	0.756	1.318**	1.179**	1.073**
	(0.511)	(0.542)	(0.590)	(0.519)	(0.526)
Ethnic motivation	0.953**	1.090**	1.144**	1.243**	0.921**
	(0.440)	(0.463)	(0.483)	(0.604)	(0.445)
Crime	0.510***	0.509***	0.362**	0.485***	0.942***
	(0.138)	(0.143)	(0.159)	(0.138)	(0.261)
Mixed carrot and stick	0.196	0.288	0.153	0.207	0.182
	(0.315)	(0.336)	(0.379)	(0.317)	(0.320)
Territory	0.504	0.459	0.544	0.496	0.551*
	(0.322)	(0.341)	(0.386)	(0.323)	(0.323)
Religious	0.324	0.281	0.290	0.908	0.371
	(0.469)	(0.498)	(0.550)	(0.609)	(0.476)
Leftist	0.660	0.975*	1.040*	1.215*	0.827
	(0.559)	(0.587)	(0.599)	(0.663)	(0.570)
Group size	0.101	0.192	0.292	0.053	0.088
	(0.212)	(0.226)	(0.248)	(0.216)	(0.217)
Single leader	−0.658	−0.713	−0.518	−0.660	−0.688
	(0.795)	(0.845)	(0.878)	(0.808)	(0.819)
Terrorism history	0.001	0.001	0.005	0.002	0.002
	(0.004)	(0.004)	(0.006)	(0.004)	(0.004)
Battle deaths	0.001**	0.000*	0.001**	0.001**	0.001**
	(0.000)	(0.000)	(0.000)	(0.000)	(0.000)

Continued

Table 4.11 *Continued*

	Model 12	Model 13	Model 14	Model 15	Model 16
	Region Controls Included (Not Shown for Space Reasons)	Alternate Regime Type Measure: Polity2	Alternate Regime Type Measure: Cheibub et al.	Alternate *Ethnic motivation* Measure	Alternate *Crime* Measure
State terror	−0.045	0.012	0.097	0.036	0.048
	(0.155)	(0.154)	(0.184)	(0.152)	(0.154)
Democracy	0.063	0.021	0.162	0.074	0.058
	(0.067)	(0.030)	(0.394)	(0.068)	(0.069)
Per capita income	0.127	0.316	0.483**	0.292	0.313*
	(0.246)	(0.199)	(0.208)	(0.186)	(0.189)
Population (log)	0.284*	0.204*	0.169	0.221*	0.201*
	(0.150)	(0.123)	(0.129)	(0.117)	(0.121)
Constant	−8.864**	−9.049***	−10.250***	−9.225***	−8.916***
	(2.733)	(2.672)	(2.925)	(2.486)	(2.520)
N	1,240	1,178	939	1,240	1,240

Note: Standard errors in parentheses. Models include group random effects and year fixed effects. $^*p < .10, ^{**}p < .05, ^{***}p < .01$.

via the Quality of Government project. This was to reduce missing data because many of the groups we analyze are in developed countries that might not have data for all years. Model 13 uses the Polity2 measure, and Model 14 uses the Cheibub et al. measure. The coefficients on these variables are statistically insignificant, and most other results are unchanged. One exception is that *Rivals* loses statistical significance in Model 14. This is interesting, but it could be because this model has approximately 25% fewer observations than other models.

Table 4.12 summarizes the overall findings from the chapter, relative to the hypotheses. We find that the hypothesized relationships are supported in most models. Hypothesis 1, representing the argument that concessions lead to less terrorism, is robust in all models except those with count dependent variables. Hypotheses 2–4, which suggested that government coercion, inter-insurgent alliances, and inter-insurgent rivalries, respectively, increase terrorism, find robust support. The two hypotheses representing insurgent relationships with the broader public or community, H5 and H6, mostly

Table 4.12 Hypotheses and Their Support Across Different Models/Dependent Variables

Hypothesis	Other Key Actor	Measure	Amount of Support in Models
1	The state	Concessions	Mostly robust, but not in models with count DVs
2	The state	Coercion	Robust
3	Other insurgents	Inter-insurgent alliances	Robust
4	Other insurgents	Inter-insurgent rivalry	Robust
5	The public	Social service provision	Mostly robust, but not with some fatality DVs
6	The public	Ethnic motivation	Mostly robust, but not with fatality DVs
7	The public	Crime	Robust

DV, dependent variable.

find support, with the exception of some models with fatality dependent variables. We find this to be an interesting caveat but consistent with the argument. Overall, the hypothesized relationships generally are supported through these tests, and it seems that insurgents' relationships with the state, other militants, and the broader community are important for helping us understand civilian targeting by insurgent groups.

4.4. Conclusion

This chapter used the BAAD2 data to test hypotheses about insurgent embeddedness and insurgent terrorism. We found substantial and robust evidence to suggest that insurgents' interactions with the state, their fellow insurgents, and the broader public help explain subsequent insurgent civilian targeting. These results are mostly consistent across multiple modeling strategies, including alternate dependent variables and the use of a lagged dependent variable. We do not find support for alternate explanations—group weakness and operating in a democratic state. The next three chapters build on these results to explore the application of the embeddedness argument to three specific types of civilian targeting: attacks focused on the general public, attacks against schools, and attacks against journalists.

References

Beck, Nathaniel. "Time-Series–Cross-Section Data: What Have We Learned in the Past Few Years?" *Annual Review of Political Science* 4, no. 1 (2001): 271–93.

Bell, Andrew, and Kelvyn Jones. "Explaining Fixed Effects: Random Effects Modeling of Time-Series Cross-Sectional and Panel Data." *Political Science Research and Methods* 3, no. 1 (2015): 133–53.

Berman, Eli, and David D. Laitin. "Religion, Terrorism and Public Goods: Testing the Club Model." *Journal of Public Economics* 92, nos. 10–11 (2008): 1942–67.

Bhavnani, Ravi, Dan Miodownik, and Hyun Jin Choi. "Three Two Tango: Territorial Control and Selective Violence in Israel, the West Bank, and Gaza." *Journal of Conflict Resolution* 55, no. 1 (2011): 133–58.

Brandt, Patrick T., John T. Williams, Benjamin O. Fordham, and B. Pollins. "Dynamic Modeling for Persistent Event-Count Time Series." *American Journal of Political Science* 44, no. 4 (2000): 823–43.

Chenoweth, Erica. "Terrorism and Democracy." *Annual Review of Political Science* 16 (2013): 355–78.

De la Calle, Luis. " Compliance vs. Constraints: A Theory of Rebel Targeting in Civil War." *Journal of Peace Research* 54, no. 3 (2017): 427–41.

Fearon, James D., and David D. Laitin. "Ethnicity, Insurgency, and Civil War." *American Political Science Review* 97, no. 1 (2003): 75–90.

Hultman, Lisa. "Attacks on Civilians in Civil War: Targeting the Achilles Heel of Democratic Governments." *International Interactions* 38, no. 2 (2012): 164–81.

Ishiyama, John, Felipe Carlos Betancourt Higareda, Amalia Pulido, and Bernardo Almaraz. "What Are the Effects of Large-Scale Violence on Social and Institutional Trust? Using the Civil War Literature to Understand the Case of Mexico, 2006–2012." *Civil Wars* 20, no. 1 (2018): 1–23.

Juergensmeyer, Mark. *Terror in the Mind of God*. Berkeley, CA: University of California Press, 2000.

Kalyvas, Stathis N. *The Logic of Violence in Civil War*. Cambridge, UK: Cambridge University Press, 2006.

Lee, Chia-yi. "Democracy, Civil Liberties, and Hostage-Taking Terrorism." *Journal of Peace Research* 50, no. 2 (2013): 235–48.

Li, Quan. "Does Democracy Promote or Reduce Transnational Terrorist Incidents?" *Journal of Conflict Resolution* 49, no. 2 (2005): 278–97.

Raleigh, Clionadh. "Violence Against Civilians: A Disaggregated Analysis." *International Interactions* 38, no. 4 (2012): 462–81.

Savun, Burcu, and Christian Gineste. "From Protection to Persecution: Threat Environment and Refugee Scapegoating." *Journal of Peace Research* 56, no. 1 (2019): 88–102.

Stanton, Jessica A. "Terrorism in the Context of Civil War." *Journal of Politics* 75, no. 4 (2013): 1009–22.

Wood, Reed M. "Rebel Capability and Strategic Violence Against Civilians." *Journal of Peace Research* 47, no. 5 (2010): 601–14.

Wood, Reed M. "From Loss to Looting? Battlefield Costs and Rebel Incentives for Violence." *International Organization* 68, no. 4 (2014): 979–99.

SECTION II
EMPIRICAL EXTENSIONS
Types of Civilian Targeting

5

Why Do Some Insurgent Groups Mostly Attack the General Public?

5.1. Introduction

Civilian victimization involves a broad range of targets. Some insurgent groups mostly attack government workers and offices. Others focus on religious leaders and institutions. The Lord's Resistance Army (LRA), however, usually attacks seemingly random civilians—the general public. In 2008, for example, the Global Terrorism Database (GTD) records 15 LRA attacks, 14 of which were against the general public.[1] In Uganda and neighboring countries, the LRA attacked villages to loot, but it killed people in the process. The LRA also killed children who resisted being kidnapped—to be soldiers or sex slaves. In December 2008, LRA members massacred hundreds of villagers in revenge attacks after a U.S.-supported military operation tried unsuccessfully to kill the group's leadership (Gettleman and Schmitt 2009). The LRA is not alone in focusing violence on the general public. Dozens of groups in our sample have years in which they mostly attack general public targets, choosing to bomb crowded markets, for example, instead of city council offices.

Insurgent attacks on the general public—seemingly random private citizens, as opposed to government officials or leaders of any type—are a significant and some say defining type of terrorism. Many definitions of terrorism emphasize that victims should be civilians or neutrals (Schmid and Jongman 2005). Others explicitly mention that a purpose of terrorism is to strike fear in the "general public." For example, the United Nations General Assembly (1995) defined terrorism, in part, as "criminal acts intended or calculated to provoke a state of terror in the general public, a group of persons or particular persons." It is also common for terrorism definitions to mention the random

---

[1] These are GTD attacks with the target type "private citizens and property." This category is discussed more later.

Insurgent Terrorism. Victor Asal, Brian J. Phillips, and R. Karl Rethemeyer, Oxford University Press. © University of Maryland National Consortium for the Study of Terrorism and Responses to Terrorism (START) 2022.
DOI: 10.1093/oso/9780197607015.003.0005

or impersonal nature of the violence (Schinkel 2009, 182; United Nations General Assembly 1995; Walzer 1977, 197). An attack on "a random person," as opposed to a soldier, diplomat, or bureaucrat, suggests the insurgents' fight is not only with the government. The violence is even more "random" when the victims are not symbols of groups possibly related to the conflict or with important public roles, such as members of rival militant groups, corporate executives, or religious leaders. Attacks on seemingly arbitrary targets are especially terrifying because they make many other citizens think that the same violence could soon happen to them (Price 1977, 52).

This chapter offers another extension of the theory and empirical tests from the previous chapter, studying possible links between insurgent embeddedness and attacks against what we call the general public. The chapter is comparable to Chapters 6 and 7 because the general public is a category comparable to news media or education targets. However, education facilities are often literally government institutions. The news media, on the other hand, is a quite specific type of target. Some news agencies are state owned, but regardless, attacks on any journalism targets have narrow goals such as literally "killing the messenger" that are a unique type of terrorism. The general public and their property, however, are more likely to be attacked to send the message to members of the population in general that they could be victims as well. These attacks also serve to warn the government that it is unable to protect its citizens. Beyond these strategic or symbolic reasons to attack seemingly random citizens, attacks on the general public could simply be an indication of group weakness—an inability to attack the government or other important institutions.

Despite the theoretical importance of attacks on the general public, there does not seem to be any empirical research on this precise subtype of terrorism. This is surprising because attacks on these types of targets are considered an ideal type of terrorism (Schinkel 2009, 182). Empirically, this type of civilian victimization is relatively common. Approximately one-third of the attacks in our data are attacks on members of the general public. More broadly, in the GTD, nearly 30% of the attacks from 1970 to 2018—more than 50,000—were classified as having "private citizens or property" as the target type.[2]

---

[2] Of the 191,464 attacks during these years in the GTD, 52,314 had such a target type. By comparison, 57,664 were against military or police targets (these are excluded from the analysis in this book), and 26,364 were against what the GTD calls "government (general)" or "government (diplomatic)."

The next section discusses in more detail what we mean by the general public and its importance in the literature on insurgency and terrorism. Then, Section 5.3 applies the theory of insurgent embeddedness to the outcome of interest and provides some expectations for how insurgent relationships are likely to help us understand why some groups mostly attack the general public. Section 5.4 provides empirical tests and their results. Section 5.5 summarizes the findings and compares them to those of models on terrorism generally and attacks against journalists or scholars. It concludes with suggestions for related future research.

5.2. The General Public, Randomness, and Insurgent Attacks

The notion of "attacks on civilians" incorporates a broad range of possible victims. As discussed previously, civilians are all those who are not members of the military or other security forces, including the police. In this chapter, we emphasize another distinction: Attacks on security targets are usually instrumental, whereas attacks on civilian targets are more symbolic (Crenshaw 1981, 379). Violence against the military has the goal of physical destruction or attrition of the enemy's fighting capacity, whereas violence against civilians often has the goal of psychological coercion of a wider audience (Merari 1993). However, there is a great deal of heterogeneity among "civilians."

We define attacks on the general public as those against civilians who do not clearly fit into a more specific category of civilian victims, such as government workers, or those viewed as representatives of business, religion, or other prominent entities.[3] Examples of attacks on the general public include an Islamic State fatal shooting of a family of 12 in Salahudin, Iraq (Xinhua 2018); a Boko Haram suicide bombing that killed 27 civilians in a market in Borno, Nigeria (PanaPress 2017); and a Real Irish Republican Army car bombing in the town center of Banbridge that wounded 35 people (Tonge 2004). To better understand attacks on the general public, we first discuss a few other categories of insurgent victims to demonstrate what distinguishes them from the general public.

---

[3] This discussion is mostly about individuals, but of course property can also be classified as belonging to the government, news media, private citizens, or other categories of potential victims. An attack on such property is classified as being an attack on that particular type of target. For simplicity, the discussion focuses on individuals, but all arguments apply to property as well.

Insurgent groups frequently target government employees and property—these qualify as civilians as well. Some government-employed civilians have direct connections to the security sector, such as politicians giving orders to the military. Other government workers are more difficult to connect to security forces, such as clerks in various bureaucracies or water treatment facility workers. However, to insurgents, these are all justified attacks on "the government." Consistent with this, the literature often groups all government employees—despite their civilian status—with security forces in terms of target types (Polo 2020). Violence against any government target is a symbolic blow against the state. As a result, a particular logic underlies this kind of violence, as the group attacks the state as a means of communicating with it. Attacks on government civilian targets are among the most conflict-related of possible civilian targets because these civilians are working for the state—against which the group is fighting. It is probably not surprising that a great deal of civilian victimization by militant groups victimizes government civilians in particular. Given that government employees can symbolize the state, we view attacks on them as distinct from attacks on non-governmental civilians that do not hold other symbolic roles—what we describe as the general public.

Insurgents attack other types of civilian targets, such as journalists (see Chapter 7) or business leaders. This violence can be symbolic as well—for example, to send a message to other journalists or to demonstrate the group's opposition to capitalism. Insurgents sometimes attack religious leaders if they represent a faith that differs from that of the insurgents. These three types of civilian victims represent specific types of targets—we argue relatively elite targets. The logic behind assaults on these types of victims is likely to be different from that of attacks on the general public. For these reasons, we exclude journalists and other prominent categories of civilians from our definition of "general public."

A key element of attacks on the general public is a perceived *randomness* to the violence. According to many scholars, randomness is an important aspect of the overlapping concept of terrorism. Walzer (1977, 197), in his classic book on morality and war, argues that "randomness is the crucial feature of terrorist activity." Schinkel (2009, 182) argues that "randomness of immediate or direct targets is a feature of the most ideal-typical forms of terrorism." In reality, a great deal of terrorism or civilian targeting is not random. Militants attack specific politicians, symbolic institutions, and even time attacks to occur on, or avoid, particular holidays (Reese, Ruby, and Pape

2017).[4] However, consistent with Schinkel's argument, there is an ideal type of civilian victimization that does seem to be random. Many attacks target members of the general public, in that sense random people, for several reasons.

Insurgents attack members of the general public to send a message to the broader public and/or the government (Kydd and Walter 2006). Regarding the broader public, a fundamental aspect of attacks on members of the general public is that because of the randomness of victims, other civilians might reasonably think that such violence could subsequently victimize themselves as well. Price (1977, 52) argues that with terrorism in particular, beyond the victims of terrorism, there is the broader audience—the group who identifies with the victim and who "receive[s] the implicit message, 'You may be next.'" When the victims seem to be targeted at random—they are not targeted because they are important officials or work at an institution with symbolic value—then all members of the public might reasonably fear they could be next. A more specific logic is that insurgents attack civilians in the general public when they suspect individuals of collaborating with the government or rival groups—or simply to coerce civilians to join their side (Merari 1993). Merari, quoting Horne (1977), notes that during the Algerian civil war, at least in the initial years, the National Liberation Front killed far more Algerians than Europeans, and often in gruesome manner to maximize the shock effect. Attacks such as these serve as a message to all other people who might cooperate with the enemy or even remain neutral.

In addition to such attacks striking fear into other members of the general public, they also send a message to the government. Insurgents often attack civilians to show the state that it is unable to protect its citizens and to coerce or extract concessions from the government (Hultman 2012). When civilians are attacked, governments, especially democratic governments, feel pressured to respond, often giving in to the militants' demands (Stanton 2013; Thomas 2014). Overall, Kydd and Walter (2006, 66) summarize why militants might attack seemingly random civilians in a strategy of intimidation: The idea is to show that insurgents "have the power to punish who disobeys them, and that the government is powerless to stop them."

---

[4] Research on Israel finds that terror victims are not randomly targeted in the sense of casualties being equally distributed across subpopulations. Terror attacks seem to target specific groups (Canetti-Nisim, Mesch, and Pedahzur 2006).

Empirically, attacks on the general public are a fairly common tactic for insurgent groups. As noted previously, approximately one-third of all attacks by groups in the Big, Allied, and Dangerous II Insurgency data are attacks on the general public. Groups that attack civilians are likely to also target the general public—at least some of the time. However, there is substantial variation across insurgent groups and across time regarding the extent that groups attack the general public compared to other targets. For some groups, such as the Armed Islamic Group, al-Aqsa Martyrs Brigade, and the LRA, attacks on the general public comprise the majority of their attacks.[5] However, most insurgent groups attack the general public at lower rates. For example, the Baloch Liberation Army, Boko Haram (perhaps surprisingly), and the Free Syrian Army only victimized the general public in a minority of their attacks, at least during the years of our data (1998–2012). These groups were more likely to attack other types of targets, such as government, education, or religious targets. The next section offers arguments and related hypotheses on what explains this variation.

5.3. Why Groups Focus Their Attacks on the General Public

There are many explanations of civilian victimization, but a particular set of arguments are necessary to account for why some insurgent groups are likely to mostly target the general public—as opposed to government workers or other common types of civilian targets. Some of the factors included in the insurgent embeddedness argument are likely to help explain attacks on civilians. There are also likely to be some differences, based on the logic of why groups attack civilians (especially as opposed to government or other targets), as well as groups' relationships with the broader public.

In the next sections, we outline hypotheses, some of which overlap with those in Chapter 2. This chapter's hypotheses are summarized in Table 5.1 for readers' convenience. We expect the same relationships between some factors and attacks on the general public as we do between those factors and terrorism generally. However, there are also expected differences. Hypotheses 3 and 5 of this chapter are different from those of Chapter 2 in that we do not expect a relationship between inter-group alliances or social service

---

[5] See the Research Design section for specific measurement.

Table 5.1 General Public Hypotheses

Hypothesis	Key Independent Variable Concept	Expected Impact on General Public Attacks
1	Concessions	Decreased likelihood
2	Coercion	Increased likelihood
3	Intergroup alliances	No relationship
4	Intergroup rivalry	Increased likelihood
5	Social service provision	No relationship
6	Ethnic motivation	Increased likelihood
7	Crime	Increased likelihood

provision and attacks on the general public. The hypotheses are discussed in more detail below.

5.3.1. The Role of the State

Chapter 2 theorized that government carrots or sticks should be associated with subsequent insurgent civilian victimization. Regarding carrots, or concessions from the state to the insurgent organization, it was argued that concessions are likely to lead to reduced civilian targeting. We suspect that this relationship is likely to hold regarding attacks on the general public. When governments provide insurgents concessions, this is generally in the hopes that the insurgents will stop their violence. This should especially be the case for attacks on the public. Analysis in Chapter 4 showed that government concessions are associated with a lower likelihood of attacks on civilians. Consistent with this, groups that have received concessions should be less likely to focus attacks on the general public. Terrorizing the general public should especially concern governments because the state is supposed to protect its citizens. Such an attack profile would make the government unlikely to continue with concessions. Thus, insurgents should anticipate this and be less likely to target the general public after having received concessions.

General public H1: Concessions to insurgent groups are associated with a *lower* likelihood of attacks focused on the general public.

Repressive or coercive actions against insurgent groups are likely to lead to the insurgents targeting the general public. Previous chapters explained why "sticks" can lead to a backlash effect. Insurgents want to respond by punishing the state, but the coercion might have weakened them, making the group less able to attack harder targets such as the military. The relationship should be similar regarding attacks on the general public. If government pressure on insurgents makes them respond with violence against civilians, insurgents might slightly prefer to attack government civilian targets, such as bureaucrats or random government property, to more directly signal their anger about the coercion, and implicitly or explicitly threaten more such violence if coercion continues. Yet government civilian targets, although "soft," can still be relatively protected (e.g., with security guards) compared to random members of the public. In addition, governments are concerned about attacks on their citizens, and militant groups exploit this by attacking civilians to intimidate the government (Kydd and Walter 2006). Thus, if coercion often encourages insurgents to target civilians, it should similarly do so with the vulnerable subcategory of the general public.

General public H2: Coercion toward insurgent groups is associated with a *higher* likelihood of attacks focused on the general public.

5.3.2. Insurgent Relationships and Attacks on the General Public

Cooperation or alliances among insurgent groups has been shown to lead to increasingly more deadly terrorism (Asal and Rethemeyer 2008; Horowitz and Potter 2014). Yet it is not clear that it should cause insurgents to focus their attacks on the general public in particular. The main causal mechanisms argued in Chapter 3 to explain the association between alliances and civilian targets are resource aggregation (e.g., Asal and Rethemeyer 2008) and tactical diffusion or learning new tactics (Horowitz 2010; Jackson et al. 2005, 44; Kettle and Mumford 2017). Regarding resource aggregation, alliances can help groups more effectively carry out particularly lethal attacks and also attacks against harder targets—attacks that require more resources. However, attacks on the general public are generally soft-target attacks not requiring an unusual amount of capacity. So it is not clear that alliances should be associated with a greater likelihood of attacks on the general public.

Regarding tactical diffusion or learning through insurgent group alliances, this is especially important for particular tactics or newly learned ones like attacks on schools (see Chapter 6) or aviation targets. Attacks on the general public generally require less sophistication, and the general public overall is not a "new" type of target. There are some tactics, such as suicide bombing, that could be learned from allies and thus have an impact on an insurgent group's propensity to attack the general public. However, despite some high-profile examples, suicide attacks are most often used against hard targets (Piazza 2020). Suicide attacks are expensive, and they are helpful for penetrating secure perimeters or otherwise difficult-to-access targets. This suggests they are more often used against, for example, government buildings. As a result, the general public is less likely to be affected by them. In summary, in contrast with the book's primary hypothesis that alliances should lead to more civilian targeting generally, it is not obvious that insurgent alliances should affect the probability of groups focusing their attacks on the general public.

> **General public H3**: Intergroup alliances are *not* associated with attacks focused on the general public.

In contrast with alliances, intergroup rivalry should probably foment attacks on the general public. There are two principal causal mechanisms linking rivalry with civilian targeting: outbidding and enmity toward those associated with the rival. Through both of these mechanisms, adversarial relationships are likely to inflict a heavy toll on the general public. Outbidding encourages more attacks and more severe types of attacks against civilians (Bloom 2005; Conrad and Greene 2015). This could lead to a higher proportion of attacks targeting the general public in particular. Attacks on seemingly random citizens are relatively easy to carry out, so when groups want to increase their quantity of attacks, we should probably expect more of these kinds of attacks. One possible argument against the outbidding mechanism leading to bloodshed against the general public is that the point of outbidding is to gain more public support—thus targeting the broader public seems counterproductive. However, insurgent groups do not necessarily want the support of the entire public but, rather, a particular segment of the public for which they are competing with rival groups. Examples include the members of the broader ethnic group they are trying to represent or members of the broader ideology (for instance, the left) that they seek to win over. Attacks on the general public are not random in this context, but they are against civilians not in their key demographic.

Relatedly, rivalry often leads to violence against civilians believed to be associated with the rival. Just as insurgents attack civilians who support the state, they also attack members of the general public who seem to be a support base for their insurgent rival(s). This can occur related to outbidding or completely independently from it. Examples include ISIS attacks on Shia communities as it battled Shia insurgents, the Taliban attacking Uzbek communities as it fought Uzbek militants, and Hamas violence against perceived Fatah supporters as the two groups targeted each other. Note that in some cases, it might not be that intergroup rivalry precedes public targeting—it could be that both phenomena are caused by a broader ethnic or other dispute. Regardless, through these multiple pathways, insurgent rivalry should be associated with insurgent groups focusing their attacks more on the general public.

> **General public H4**: Intergroup rivalries are associated with a *higher* likelihood of attacks focused on the general public.

5.3.3. Public Interactions: Social Services, Ethnic Motivations, and Crime

The final hypotheses of Chapter 3 proposed that violence against civilians is associated with how insurgent groups interact with the public. The hypothesized factors—social service provision, ethnic motivations, and crime—should have important but sometimes distinct relationships with attacks on the general public. Regarding social service provision, this type of activity generally improves group capacity because it attracts recruits and donations. This can lead to more violent activity (Flanigan 2008; Iannaccone and Berman 2006). In previous chapters, we argued that this should also apply to violence against civilians in general. However, regarding attacks directed at the general public, it is less clear. Groups that provide social services, and reap the benefits in terms of mobilization, do seem to attack more civilians (see Chapter 4). This violence, however, could be directed more at government or other types of targets instead of the general public. Insurgent groups that provide social services are trying to get the public, or at least a substantial portion of it, on their side. As a result, it would seem counterintuitive for these same insurgent groups to attack random civilians more than other types of civilian targets.

In addition, it has been argued that social service provision makes insurgent groups stronger, more capable of carrying out complex violence (Berman and Laitin 2008). These groups, then, should be more likely to attack targets such as security forces and highly symbolic targets such as government buildings. Some of these groups, such as Hamas, do use their substantial power to attack the public. However, other insurgent organizations that provide social services do not use most of their violence against the general public. They are able to attack more specific and potentially valuable targets, so they probably do so. Anecdotally, a number of groups that provide social services seem to fit this pattern. The Kurdistan Workers Party (PKK), Revolutionary Armed Forces of Colombia (FARC), and even al-Qaeda in the Arabian Peninsula (AQAP) target the general public less than they do government and other kinds of targets. Given the somewhat mixed expectations—more ability to use violence including attacking the general public, but also reasons to refrain from such attacks—we suggest the following hypothesis:

General public H5: Social service provision is *not* associated with attacks focused on the general public.

We have argued that ethnic motivations suggest a particular relationship between insurgent organizations and the wider public. As insurgent groups seek to represent an ethnic community, they might try to coerce members of the community into supporting them. This has often been the situation with the Taliban and Pashtuns in Afghanistan (Felbab-Brown, Trinkunas, and Hamid 2017). In addition, ethnically motivated insurgent organizations might attack anyone who does not belong to the ethnic group they seek to represent. In general, sectarian groups are more likely to attack soft targets than hard or official targets (Polo and Gleditsch 2016). An example of such sectarian violence can be seen in the National Democratic Front of Bodoland, an insurgent group in northeast India representing the Bodo ethnicity. This group has attacked civilians from the Adivasi ethnic group, among others (*The Guardian* 2014). Both of these dynamics—coercing their own population and attacking members of other ethnic groups—seem relevant for understanding attacks on the general public. As a result, it seems likely that insurgent groups with ethnic goals should frequently attack the general public, as opposed to other types of civilian targets.

General public H6: Ethnic motivations are associated with a *higher* likelihood of attacks focused on the general public.

Finally, involvement in crime such as drug trafficking or kidnapping is associated with civilian victimization (see Chapter 4), and it probably should lead to attacks on the general public as well. Crime involvement suggests a predatory relationship with the general public, so insurgent groups engaging in crime should show little restraint regarding such attacks. In addition, the involvement in crime could directly lead to violence against the general public, as insurgents kill kidnapping victims whose families do not deliver a ransom or attack extortion victims who do not pay—perhaps publicly to send a message to others. For example, the Indian secessionist group All-Tripura Tiger Force is substantially involved in kidnapping and extortion, and it has injured and killed many members of the general public through these activities (*The Economic Times* 2018). According to our data, the group attacks the general public more than it does any other target type, including government targets. Other groups are likely to engage in similar patterns. This suggests the following hypothesis:

General public H7: Involvement in crime is associated with a *higher* likelihood of attacks focused on the general public.

5.4. Data and Analysis on Attacks Against the General Public

This section presents a replication of the empirical models in Chapter 4, with the alternate and more specific dependent variable indicating groups that mostly attack the general public. Because the data and models were described in Chapter 3 and 4, this information is not repeated. The dependent variable for this chapter, *Attacks focused on the general public*, derives from information in the GTD. The GTD includes a "target type" variable. Each attack is classified as being one of 22 categories, such as government (general), government (diplomatic), religious figures/institutions, educational institution (see Chapter 6), or journalists & media (see Chapter 7). The key category for this chapter is *Private citizens and property*. The GTD defines the variable as follows: "This value includes attacks on individuals, the public in general or attacks in public areas including markets, commercial streets, busy intersections and pedestrian malls" (START 2019). The codebook further specifies that violence against weddings and funerals is also included, as are attacks against students, as long as the attacks are not "carried out in an education setting" (START 2019).

As in Chapter 4, we use a dichotomous dependent variable because we are more interested in an indicator of group behavior than the specific number of attacks a group uses. However, *Attacks focused on the general public* is measured somewhat differently than the other dependent variables in this book. We operationalize the concept with a variable coded "1" if the *majority* of the group's attacks in a given year were against "private citizens and property." This measurement is helpful because most groups that attack civilians also attack the general public *some* of the time. An alternate measure—indicating any attacks on the general public—is also highly correlated with the dichotomous measure of attacks against civilians. Instead, our measure focuses on instances when *most* attacks by a group are on the general public. Our more nuanced measure is helpful because it identifies years when insurgent groups seem focused on the particular and important "ideal type" of terrorism (according to some others) of attacks against the general public.[6]

Figure 5.1 shows the distribution of attacks on the general public, both by the number of groups that mostly carry out these kinds of attacks (left) and by the simple count of such attacks over time (right). The number of attacks on the general public increases greatly in the 2000s, consistent with attacks on civilians generally. However, regarding our more specific dependent variable, when groups mostly target the general public in a particular year, this shows a different kind of variation. The number of groups focused on attacking the general public in a given year ranges from 6 to 18, with the average at just over 11. Nearly 10% of the groups in the sample are coded as conducting this type of violence each year. It is important to note that it is not the same small set of groups each year coded for this variable. Nearly half of the groups in our data (62) are coded as mostly attacking the general public

---

[6] A measure indicating that a majority of a group's attacks were on schools or the news media is not possible because these two types of attacks are almost never the majority of groups' attacks. Violence against the general public is much more common.

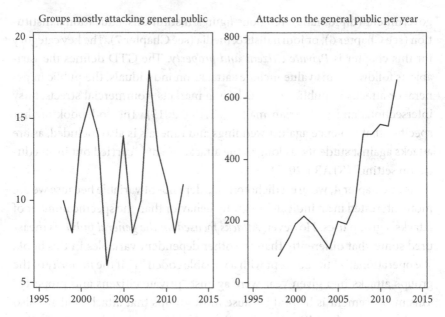

Figure 5.1   Insurgent groups' attacks on the general public over time.

in at least 1 year of the sample. This is a fairly common and important to understand attack profile.

5.4.1.   Results

Table 5.2 replicates the main models of Chapter 4 (Table 4.8), replacing the dependent variable with *Attacks focused on the general public*. Model 1 includes only control variables, Model 2 contains only the key independent variables, and Model 3 is the full or main model. The results have some interesting differences compared to those of Table 4.8. Of the variables representing hypothesized relationships, *Carrot* is statistically insignificant. This is unexpected because government concessions were associated with civilian targeting generally and attacks against schools. It could be that government concessions cause insurgent groups to reduce their attacks on government targets—politicians, government buildings, and other symbols—but not necessarily the general public. This suggests unique logics behind each type of civilian targeting.

Other variables have results that mirror those of the models of civilian targeting (see Chapter 4). *Stick, Rivalry,* and *Ethnic motivation* are all

Table 5.2 Logit Models of Attacks Focused on the General Public

	Model 1	Model 2	Model 3
	Only Control Variables	Only Key Variables	Main Model
Carrot		−0.591	−0.476
		(0.744)	(0.780)
Stick		0.719**	0.774**
		(0.255)	(0.347)
Alliances		0.106**	0.054
		(0.052)	(0.057)
Rivalry		0.747**	0.520**
		(0.227)	(0.240)
Social services		0.279	0.206
		(0.459)	(0.467)
Ethnic motivation		0.758*	0.746*
		(0.400)	(0.441)
Crime		0.210	0.200
		(0.132)	(0.140)
Mixed carrot and stick	−0.284		0.095
	(0.285)		(0.412)
Territory	0.022		−0.132
	(0.338)		(0.373)
Religious	0.610		0.505
	(0.377)		(0.481)
Leftist	−0.572		−0.199
	(0.476)		(0.561)
Group size	0.115		0.132
	(0.217)		(0.234)
Single leader	−1.608		−1.378
	(0.985)		(1.051)
Private target history	0.001		−0.000
	(0.004)		(0.004)
Conflict battle deaths	0.000		0.000
	(0.000)		(0.000)
State terror	0.353**		0.190
	(0.171)		(0.184)
Democracy	0.244***		0.160**
	(0.067)		(0.075)

Continued

Table 5.2 *Continued*

	Model 1	Model 2	Model 3
	Only Control Variables	Only Key Variables	Main Model
GDP per capita	−0.053		−0.148
	(0.176)		(0.190)
Population (log)	−0.102		−0.114
	(0.105)		(0.118)
Constant	−3.868	−4.355***	−3.303
	(2.407)	(0.604)	(2.643)
N	1,240	1,240	1,240

Note: Standard errors in parentheses. Models include group random effects and year fixed effects. *p < .10, **p < .05, ***p < .01. All independent variables are lagged 1 year. This table replicates Table 4.8, in which *Terrorism* was the dependent variable.

statistically significant and positively signed.[7] This is consistent with expectations, and it suggests overlaps between civilian targeting and the more specific phenomenon of focusing on the general public. When insurgent groups are the subject of government coercion, they have a rival, or when they have ethnic motivations, they are likely to focus their attacks mostly on the general public. This suggests the importance of these three factors for explaining civilian targeting of several types. Regarding substantive significance, as in Chapter 4, we show marginal effects in Figure 5.2. *Ethnic motivation* and *Stick* have the largest estimated effect of any variables in the model—a 6 percentage-point increase in the probability of a group mostly focusing on the general public. The effect is a 4 percentage-point increase for *Rivalry*.

Regarding Hypothesis 3 in this Chapter, which suggests no relationship with alliances, this hypothesis finds support. The coefficient on *Alliances* is statistically insignificant. This is consistent with the causal mechanisms we outlined, resource aggregation and tactical diffusion, not necessarily being relevant for attacks on the general public. Another variable, *Social services*, is statistically insignificant as expected. This is also suggestive of nuance with causal mechanisms. We argued that social services help groups become more

[7] The coefficient on *Ethnic motivation* is only statistically significant at the 90% level. Ethnically motivated insurgent groups do seem likely to attack the civilians of other ethnicities, and sometimes their own, but perhaps concern about winning over their own wider ethnic groups dampens their propensity to attack the general public.

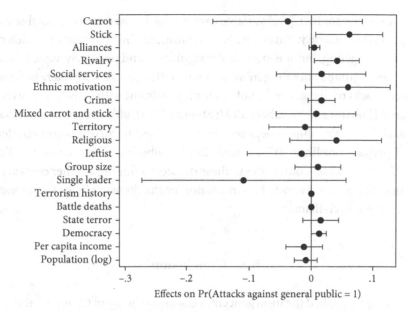

Figure 5.2 Substantive significance of variables in the main model. Average marginal effects are shown with 95% confidence intervals.

powerful and more lethal—but not necessarily against the general public. Because groups that provide social services are trying to get the general public on their side, they do not seem to be especially likely to attack the general public, and the (non)results support this.

One other hypothesized variable is statistically insignificant, but not one we expected to be. It is unclear why *Crime* would not be related to focusing attacks on the general public, as there are reasons to expect crime to be related to civilian victimization of various types. However, it is plausible that some groups engaged in crime need the public on their side—to inform the group about police patrols, for example, and to *not* inform on the group to the government. Although there are instances of insurgent groups engaging in crime and attacking seemingly random civilians, there are also cases of groups such as the PKK or FARC that participate in illicit business for profit and seem to try to minimize attacks on the general public. Further research should probe the relationships between crime involvement and civilian victimization to better understand causal pathways and conditional relationships.

Finally, the results of the control variables in Table 5.2 are mostly similar to those in Table 4.8 (the main results of Chapter 4). Interesting group-level variables such as territory are not related to attacking the general public, and most state or conflict-level variables are not related as well. One notable exception is *Democracy*. This variable was statistically insignificant in models of civilian targeting, but it is statistically significant and positively signed here. Insurgent groups operating in more democratic countries are likely to focus their attacks on the general public. This is consistent with some previous research (Eubank and Weinberg 1994; Stanton 2013), although other work has noted the relationship(s) between democracy and terrorism is complicated and perhaps nonlinear (Chenoweth 2013; Gaibulloev, Piazza, and Sandler 2017). This chapter contributes to these debates by finding another context in which democracy seems to be a motivator for attacks on civilians—or at least some types of civilians.

5.5. Conclusion

This chapter applied the insurgent embeddedness theory of Chapter 2 and the empirical modeling of Chapter 4 to an important and related research question: Why do some insurgent groups mostly target the general public? The question is important because attacks on all civilians—the main dependent variable in Chapter 4—include many government victims such as politicians as well as individuals representing very specific targets, such as business or religious leaders. Limiting the dependent variable to only the general public captures the element of "randomness" that is important for many definitions of terrorism (Schinkel 2009, 182; Walzer 1977, 197). It is also consistent with the notion of terrorism striking fear into all citizens, making them believe they could be next (Price 1977, 52). Attacks on the general public are one of the most common types of terrorism, and nearly half of the groups in our sample have at least 1 year in which they mostly attack the general public.

Analyses demonstrated that many variables from our embeddedness argument help explain this extreme type of civilian targeting. Consistent with expectations, government coercion, intergroup rivalry, and ethnic motivation are all associated with groups focusing their attacks on the general public. Other key group attributes—intergroup alliances and social service provision—were not related to groups mostly attacking the general public, as we expected. Intergroup alliances have been argued to lead to civilian targeting through resource aggregation and tactical diffusion (Asal and Rethemeyer 2008; Horowitz 2010), and we did not think these causal mechanisms should indicate a propensity to focus attacks on the general public. Government and harder targets should be more likely. Similarly, it makes sense that social service provision, although often associated with militant group strength and lethality, should not necessarily lead to attacks on the general public. Insurgent groups providing social services want the public, or at least a substantial portion of it, on their side.

A surprising finding was that involvement in crime is not associated with attacking the general public. One might think that groups engaging in illicit fundraising would not be too concerned about public support, and they might even attack the public as a part of their criminal behavior, such as kidnapping or extortion. However, we noted the examples of the PKK and FARC, which engage in crime and usually victimize the public at lower rates than other groups. Future research should examine the possible links between crime involvement and civilian victimization in more detail. Another interesting finding was for a control variable, democracy. Insurgent groups in more democratic countries are more likely to focus attacks on the general public. This is consistent with some work on terrorism (Eubank and Weinberg 1994; Stanton 2013), although possible relationships with regime type are one of the most debated subjects in terrorism studies (Chenoweth 2013). Some prominent scholars have dismissed the notion of a linear relationship between regime type and civilian victimization (Gaibulloev et al. 2017), but the relationship might depend on what kind of civilians are victimized. This, too, is worthy of future research.

This chapter begins our extension of the main empirical tests (see Chapter 4), applying the theoretical model to three related dependent variables. Chapter 6 examines how the model holds up when applied to a related outcome—attacks on schools.

References

Asal, Victor, and R. Karl Rethemeyer. "The Nature of the Beast: Organizational Structures and the Lethality of Terrorist Attacks." *The Journal of Politics* 70, no. 2 (2008): 437–49.

Berman, Eli, and David D. Laitin. "Religion, Terrorism and Public Goods: Testing the Club Model." *Journal of Public Economics* 92, nos. 10–11 (2008): 1942–67.

Bloom, Mia. *Dying to Kill: The Allure of Suicide Terror.* Columbia University Press, 2005.

Canetti-Nisim, Daphna, Gustavo Mesch, and Ami Pedahzur. "Victimization from Terrorist Attacks: Randomness or Routine Activities?" *Terrorism and Political Violence* 18, no. 4 (2006): 485–501.

Chenoweth, Erica. "Terrorism and Democracy." *Annual Review of Political Science* 16 (2013): 355–78.

Conrad, Justin, and Kevin Greene. "Competition, Differentiation, and the Severity of Terrorist Attacks." *The Journal of Politics* 77, no. 2 (2015): 546–61.

Crenshaw, Martha. "The Causes of Terrorism." *Comparative Politics* 13, no. 4 (1981): 379–99.

Eubank, William Lee, and Leonard Weinberg. "Does Democracy Encourage Terrorism?." *Terrorism and Political Violence* 6, no. 4 (1994): 417–35.

Felbab-Brown, Vanda, Harold Trinkunas, and Shadi Hamid. *Militants, Criminals, and Warlords: The Challenge of Local Governance in an Age of Disorder*. Washington, DC: Brookings Institution Press, 2017.

Flanigan, Shawn Teresa. "Nonprofit Service Provision by Insurgent Organizations: The Cases of Hizballah and the Tamil Tigers." *Studies in Conflict & Terrorism* 31, no. 6 (2008): 499–519.

Gaibulloev, Khusrav, James A. Piazza, and Todd Sandler. "Regime Types and Terrorism." *International Organization* 71, no. 3 (2017): 491–522.

Gettleman, Jeffrey, and Eric Schmitt. "U.S. Aided a Failed Plan to Rout Ugandan Rebels." *The New York Times*. February 6, 2009. https://www.nytimes.com/2009/02/07/world/africa/07congo.html.

Horne, Alistair. *A Savage War of Peace: Algeria, 1952–1962*. London, UK: Macmillan, 1977.

Horowitz, Michael C. "Nonstate Actors and the Diffusion of Innovations: The Case of Suicide Terrorism." *International Organization* 64, no. 1 (2010): 33–64.

Horowitz, Michael C., and Philip BK Potter. "Allying to Kill: Terrorist Intergroup Cooperation and the Consequences for Lethality." *Journal of Conflict Resolution* 58, no. 2 (2014): 199–225.

Hultman, Lisa. "Attacks on Civilians in Civil War: Targeting the Achilles Heel of Democratic Governments." *International Interactions* 38, no. 2 (2012): 164–81.

Iannaccone, Laurence R., and Eli Berman. "Religious Extremism: The Good, the Bad, and the Deadly." *Public Choice* 128, 1–2 (2006): 109–29.

Jackson, Brian A., John C. Baker, Kim Cragin, John Parachini, Horacio R. Trujillo, and Peter Chalk. *Aptitude for Destruction. Volume 1: Organizational Learning in Terrorist Groups and Its Implications for Combating Terrorism*. Santa Monica, CA: RAND, 2005.

Kettle, Louise, and Andrew Mumford. "Terrorist Learning: A New Analytical Framework." *Studies in Conflict & Terrorism* 40, no. 7 (2017): 523–38.

Kydd, Andrew H., and Barbara F. Walter. "The Strategies of Terrorism." *International Security* 31, no. 1 (2006): 49–80.

Merari, Ariel. "Terrorism as a Strategy of Insurgency." *Terrorism and Political Violence* 5, no. 4 (1993): 213–51.

PanaPress. "Nigeria: Boko Haram Attacks in Northeast Nigeria Leave 27 Dead, Scores Wounded." August 16, 2017.

Piazza, James A. "Suicide Attacks and Hard Targets: An Empirical Examination." *Defence and Peace Economics* 31, no. 2 (2020): 142–59.

Polo, Sara M. T. "The Quality of Terrorist Violence: Explaining the Logic of Terrorist Target Choice." *Journal of Peace Research* 57, no. 2 (2020): 235–50.

Polo, Sara M. T., and Kristian Skrede Gleditsch. "Twisting Arms and Sending Messages: Terrorist Tactics in Civil War." *Journal of Peace Research* 53, no. 6 (2016): 815–29.

Price, H. Edward. "The Strategy and Tactics of Revolutionary Terrorism." *Comparative Studies in Society and History* 19, no. 1 (1977): 52–66.

Reese, Michael J., Keven G. Ruby, and Robert A. Pape. "Days of Action or Restraint? How the Islamic Calendar Impacts Violence." *American Political Science Review* 111, no. 3 (2017): 439–59.

Schinkel, Willem. "On the Concept of Terrorism." *Contemporary Political Theory* 8, no. 2 (2009): 176–98.

Schmid, Alex P., and Albert J. Jongman. *Political Terrorism: A New Guide to Actors, Authors, Concepts, Data Bases, Theories, & Literature*. New Brunswick, NJ: Transaction, 2005.

Stanton, Jessica A. "Terrorism in the Context of Civil War." *Journal of Politics* 75, no. 4 (2013): 1009–22.

START (National Consortium for the Study of Terrorism and Responses to Terrorism). "GTD Codebook: Inclusion Criteria and Variables." October 2019. https://www.start. umd.edu/gtd/downloads/Codebook.pdf.

The Economic Times. "Government Slaps Fresh Ban on Two Insurgent Groups of Tripura." October 4, 2018. https://economictimes.indiatimes.com/news/defence/government-slaps-fresh-ban-on-two-insurgent-groups-of-tripura/articleshow/66068470.cms?utm _source=contentofinterest&utm_medium=text&utm_campaign=cppst.

The Guardian. "Bodoland Tribal Separatists Kill at Least 34 in Assam, Indian Police Say." 2014. https://www.theguardian.com/world/2014/dec/23/bodoland-tribal-separatists-kill-34-assam-indian-police-say.

Thomas, Jakana. "Rewarding Bad Behavior: How Governments Respond to Terrorism in Civil War." *American Journal of Political Science* 58, no. 4 (2014): 804–18.

Tonge, Jonathan. "'They Haven't Gone Away, You Know.' Irish Republican 'Dissidents' and the 'Armed Struggle.'" *Terrorism and Political Violence* 16, no. 3 (2004): 671–93.

United Nations General Assembly. "General Assembly Resolution 49/60: Measures to Eliminate International Terrorism." February 17, 1995. https://www.un.org/ga/search/ view_doc.asp?symbol=A/RES/49/60.

Walzer, Michael. *Just and Unjust Wars: A Moral Argument with Historical Illustrations*. New York, NY: Basic Books, 1977.

Xinhua. "12 Family Members Killed by Gunmen in Iraq." 2018. http://www.xinhuanet. com/english/2018-06/02/c_137225686.htm.

6

Why Do Some Insurgent Groups Attack Schools?

6.1. Introduction

On February 1, 2013, a huge explosion in the village of Balsillas in southern Colombia destroyed a school that served 60 students, as well as its dormitories. The Revolutionary Armed Forces of Colombia (FARC) apparently carried out the attack, angry that the U.S. embassy and Colombian Army had funded schools in the region (Montaño 2013). Fortunately, no one was injured in the bombing, one of several FARC attacks against schools. On this occasion, FARC seems to have been more interested in sending a message to the state than killing children. Other attacks on educational institutions, however, have not been so merciful. For example, in 2014, Tehrik-i-Taliban Pakistan members attacked a school for military children in Peshawar, Pakistan, killing 149 people, mostly children. Why do insurgent groups sometimes target schools or universities?

This chapter presents an extension of the main theory and empirical tests reported in Chapter 4, examining the determinants of insurgent group attacks on educational targets such as schools. On the one hand, this can be viewed as an "extreme test" of the model, examining if the relationships demonstrated in Chapter 4 also hold for an unusually vicious type of non-combatant targeting—attacking some of the most vulnerable targets. On the other hand, schools might be attacked for some unique reasons, aside from those behind terrorism generally, so it is interesting to think about the logic of targeting this particular type of institution. Overall, however, the focus on schools is consistent with the book's general motivation of seeking to understand why insurgents sometimes use terrorism.

The terrorist attacks analyzed in Chapter 4 include a broad variety of violence. Some of these attacks could be more or less terroristic than others. For example, ostensible noncombatants targeted in civilian attacks listed in the Global Terrorism Database (GTD) could be government informants or

Insurgent Terrorism. Victor Asal, Brian J. Phillips, and R. Karl Rethemeyer, Oxford University Press. © University of Maryland National Consortium for the Study of Terrorism and Responses to Terrorism (START) 2022. DOI: 10.1093/oso/9780197607015.003.0006

persons directly contributing to the state's war-making process. The victimization of individuals involved in conflict processes in these ways might be thought of as less terroristic compared to attacks on relatively random or uninvolved civilians. Attacks on schools are especially heinous or surprising because schools and students are unlikely to be involved in an armed conflict.

Educational institutions, like those engaged in charity or religion, are specifically named in Article 56 of the 1907 Hague Convention agreement as protected cultural entities (International Peace Conference 1907). Attacks against them are prohibited in international and non-international (i.e., civil) armed conflict (International Peace Conference 1907). Attacks on educational facilities, whether primary or secondary schools, are shocking because they generally target children. Tokdemir and Akcinaroglu (2016, 271) describe intentional attacks on children as "morally incomprehensible," and the authors use such attacks as an indicator of a terrorist group with a negative reputation. Even terrorism against universities, although usually not victimizing children, is alarming due to the generally pacific nature of these institutions. Education is considered a human right (United Nations 1948), and damage to educational institutions is a war crime (International Peace Conference 1907).

The next section discusses the significance of educational targets in conflict and compares them to other target types. The third section considers the logic behind why militants target educational institutions, and it presents related hypotheses. The fourth section discusses empirical tests and their results. The final section concludes with implications of the findings, and it discusses differences between these findings and those of Chapter 4.

6.2. Education as a Target Type

Several horrific attacks on schools are well known. The 2014 Peshawar Army School attack is one, but a more consequential and deadly attack was the 2004 Beslan school siege in Russia. Chechen separatists held students and staff hostage for 3 days, killing dozens. After Russian forces entered the school during a rescue attempt, more than 300 people, mostly children, ended up dead.[1] Other prominent attacks include the Boko Haram kidnapping of 276 female students from a school in Chibok, Nigeria.

---

[1] This attack is not in our data because the perpetrators were not an insurgent group—they were not involved in an armed conflict beyond this single attack, according to the Uppsala Conflict Data Program.

In addition to these high-profile attacks, many other insurgents have attacked educational institutions. Of the 140 groups in this book's data set, 44 have attacked a school or university at least once. Abu Sayyaf, for example, kidnapped and later killed an elementary school principal in the Philippines in 2009. In 2002, a bomb planted at a cafeteria in the Hebrew University of Jerusalem killed seven people and wounded dozens more. Hamas claimed responsibility for the attack. In 1999, FARC claimed a bombing at the University of Antioquia in Medellín. Beyond insurgent groups engaged in armed conflict, a variety of actors using terrorism (e.g., terrorist groups or lone actors) have attacked educational facilities. The number of attacks has increased substantially in recent decades, and they also seem to be increasingly lethal with time (Bradford and Wilson 2013; Petkova et al. 2017).

Educational institutions are, of course, but one among many types of targets insurgents might strike (Moghadam, Berger, and Beliakova 2014; Polo 2020; Santifort, Sandler, and Brandt 2013). Table 6.1 outlines certain types of insurgent targets, sorted by the extent to which they are "hard" or "soft" targets and whether they are relatively conflict-involved or conflict-uninvolved. The hard and soft target distinction is familiar in the terrorism literature because terrorism often strikes the latter (Asal et al. 2009; Brandt and Sandler 2010; Eyerman 1998; Polo and Gleditsch 2016). Our operationalization of terrorism in previous chapters excluded attacks on government security forces because these attacks are similar to violence in civil conflict and less like terrorism. In addition to the hard and soft target distinction, we include a second dimension to highlight that some targets are involved in the conflict, even if soft targets, such as government officials or off-duty soldiers. Other target types, however, are relatively uninvolved in the conflict. Some conflict-uninvolved targets are hardened, such as banks

Table 6.1 Potential Targets of Insurgent Groups

	Hard Targets	Soft Targets
Conflict-Involved	Military	Government officials
	Other security forces including police	Off-duty or unarmed military personnel
Conflict-Uninvolved (Relatively)	Banks	Hospitals
		Religious institutions
		Schools

for obvious reasons. However, for the study of terrorism in civil conflict, we are especially interested in conflict-uninvolved soft targets.

We argue that targets that are relatively conflict-uninvolved, such as schools, hospitals, and religious institutions, more closely represent the ideal type of terrorism. This is because terrorism is argued to be extranormal, its victims "neutrals" (Schmid and Jongman 2005, 6). Schmid and Jongman's classic discussion of definitions of terrorism suggests that the tactic is "exceptional" in warfare (p. 16). As a result, attacks on *some* soft targets—those that can still be viewed as part of the government's war-making apparatus—are arguably less "pure" terrorism and more a continuation of "normal" insurgency or rebellion.

Schools, medical facilities, and religious institutions are prime examples of soft targets that are generally uninvolved in the perpetration of conflict and therefore are especially unlikely to be targeted legitimately. This is basically why these three types of institutions are protected by international humanitarian law (International Peace Conference 1907). Although attacking government employees or unarmed soldiers is generally still considered terrorism, attacks on schools, hospitals, and places of worship are an extreme class. They are especially *brutal*.[2] This is consistent with Fujii's (2013) notion of extralethal violence, especially horrific interpersonal violence, in the sense that it transgresses norms and could be considered a more extreme subtype of terrorism. Attacks on institutions that cannot be plausibly linked to armed conflict are especially extreme among terrorist acts.

However, even among this class of potential targets, schools stand out. It is extremely difficult for insurgents to claim that an attack on a school is justified as part of the war effort. Hospitals and religious institutions, although wholly illegal to target, might be a gray area for insurgents. Hospitals are sometimes attacked by insurgents because they are treating soldiers or members of rival militant groups. For example, when a Russian hospital near Chechnya was bombed in 2013, the government speculated that it was targeted because Russian soldiers wounded in Chechnya were treated there (Myers 2003).

Religious institutions are sometimes attacked when religion is weaponized during ethnoreligious or sectarian conflicts. Attacks by the so-called Islamic State on Shi'ite mosques are an example of this (Reuters 2019). In general,

---

[2] Weinstein (2006, 18) uses "brutality" to refer to behaviors such as rape or amputations that go above and beyond what is necessary to send a signal about the costs of defection.

soldiers are likely to be found at hospitals or religious institutions but much less likely to be in educational institutions.[3] The fact that the majority of victims at a school would be children, among the most vulnerable subgroups of noncombatants (Petkova et al. 2017), makes educational facilities a fairly unique target type. Overall, all three of these types of institutions are soft targets that are at least generally not directly involved in the perpetration of armed conflict. But educational institutions represent an extreme case—not only an illegal and unethical target type but also one that is among the most difficult for insurgents to justify as legitimate in any way. For these reasons, we think of attacks on educational institutions as an archetypical type of terrorism by insurgent groups.

6.3. Why Attack Schools? Insurgent Embeddedness and Educational Institution Targeting

The extant literature does not tell us much about which militant groups should be especially likely to attack schools. There is apparently only one study on a similar topic—why some countries experience more terrorism against schools than other countries (Fahey and Asal 2020). The researchers, focused on state attributes, find that government repression and some other attributes such as female empowerment are associated with educational attacks. Another study surveys general trends of terrorism against schools (Petkova et al. 2017). Thinking about the organizational unit of analysis, however, it is not immediately clear why some groups target schools.

The theory of insurgent embeddedness and civilian targeting presented previously can be applied to attacks on educational institutions. However, certain modifications are necessary to take into consideration how unique schools are as potential targets. The following sections consider how the propensity for an insurgent group to target schools can be affected by the seven factors hypothesized to affect civilian targeting generally: government concessions, government coercion, intergroup alliances, intergroup rivalry, social service provision, ethnic motivation, and crime. Some relationships

---

[3] A mosque in Pakistan was bombed in 2019 apparently because a Taliban leader was generally in attendance (Mashal and Masood 2019). This occurred in the context of a rivalry with ISIS: ISIS was attacking mosques where rival group members worshipped. This type of directed-at-militants violence is much less likely at educational institutions.

should be the same for educational institutions as they are for terrorism generally. Other relationships should be stronger for educational institutions. Still other relationships should not hold for educational institutions at all.

6.3.1. Insurgents, the Targeted State, and Schools

The relationships between insurgents and the government with which they fight should have important consequences for the likelihood that the insurgents attack educational institutions. Chapter 2 drew attention to carrots and sticks, and it was argued that carrots or concessions are likely to reduce civilian targeting by insurgent groups. We think the reduced likelihood of terrorism should be even stronger with the class of terrorism targeted at education institutions.

Concessions are associated with reduced subsequent terrorism by insurgent groups. This should especially be the case regarding terrorism against educational institutions because this is an especially horrific kind of violence. Attacks on schools are so unacceptable, so outside the bounds of acceptable conflict behavior, that insurgents would know that these attacks would greatly reduce their chance of future concessions. This is in part related to the signal that attacking educational institutions sends. An attack such as this paints the perpetrator as a clear "terrorist," perhaps not even an insurgent. Once a group has branded itself in this way, or once it carries out actions that allow the government to brand it in this way, states would find it quite difficult to justify concessions toward it. There are exceptions. Thomas (2014) found that governments sometimes "reward bad behavior." However, within the types of bad behavior—the types of terrorism—the extreme category of attacking schools should be the least likely to be rewarded. This suggests the following hypothesis:

School H1: Concessions to insurgent groups are associated with a *lower* likelihood of educational institution targeting.

Government coercion is likely to have similarly strong effects on school targeting, but in the opposite direction. Chapter 2 argues that when governments crack down on insurgents, as opposed to providing concessions or passively coexisting with them, insurgents often respond with violence, including terrorism. The analyses in Chapter 4 suggest support for this argument. We further argue that coercion against insurgent groups is especially likely to lead to *extreme* subsequent terrorism, such as educational targets.

Insurgent groups by definition use violence. To build or maintain legitimacy among the public, they might prefer to attack security force targets. Indiscriminate violence is often counterproductive; insurgents would use selective violence more if they had the ability to do so (Kalyvas 2006). Consistent with this, a number of studies suggest terrorism is not effective for winning wars or achieving other strategic victories (Abrahms 2006; Fortna 2015).

Despite this, insurgents often use terrorism for various reasons (Kydd and Walter 2006). Although they might refrain from using terrorism after receiving concessions, the opposite is likely to be true when governments are actively pursuing them. If states are coercing insurgents, as opposed to being conciliatory or passively tolerating them, this reduces the reasons insurgents might have to abstain from terrorism. Taking the argument to an extreme, insurgents are especially likely to abstain from the most brutal forms of violence if the state has not recently taken actions against them. This could be a costly mistake for the insurgents because it could cause the government to change its course and start coercion. Once coercion is started, however, this lowers the potential cost for the use of terrorism. As a result, coercion should be associated with subsequent terrorism, including the targeting of educational institutions. Furthermore, the relationship between coercion and attacks on educational institutions could be especially powerful given the strong incentives for insurgents to abstain from brutal violence when facing an absence of coercion.

> **School H2**: Coercion toward insurgent groups is associated with a *higher* likelihood of educational institution targeting.

6.3.2. Intergroup Alliances, Rivalry, and Attacks on Educational Institutions

Alliances among insurgent groups are likely to be related to attacks on schools. The possible relationship between rivalries and attacks on educational institutions is not as clear. Regarding alliances, militant groups gain resources through cooperation that can help them carry out more attacks in general and also more lethal and complex attacks (Akcinaroglu 2012; Asal and Rethemeyer 2008; Horowitz and Potter 2014). Increased resources mean increased capacity, and this could be helpful for groups seeking to attack educational institutions. Attacks on schools or universities, although usually considered soft targets, are generally more complex than other types of terrorist attacks. These are often large structures, perhaps with a fence or a guard, and there are usually many people in them. This suggests a more complicated attack than one against a private home, a person walking in public, or a public bus.

In addition to alliances suggesting increased capacity to attack educational targets, alliances also encourage tactical diffusion (Horowitz and Potter 2014). Attacks on schools, such as suicide attacks, could be such an extreme and unusual tactic that insurgent groups only conduct such attacks after their allies do so. This would be consistent with the fact that attacks on schools, which had been rare until the early 2000s, dramatically increased starting in approximately 2005 (Petkova et al. 2017). Diffusion of these types of tactics seems more likely than the diffusion of terrorism itself, which is a less original or unusual approach for an insurgent group. As a result, if alliances encourage diffusion, the effects of diffusion on educational attacks should be stronger than the effects on terrorism generally.

School H3: Intergroup alliances are associated with a *higher* likelihood of educational institution targeting.

Interorganizational rivalry is widely associated with increased terrorism, but it is unclear that rivalry should encourage attacks on educational institutions in particular. On the one hand, competition broadly defined can be associated with more spectacular terrorism, as Bloom (2004, 2005) argues with outbidding and suicide bombing. This could lead to attacks on schools as a more extreme type of tactic. Interestingly, a case study of the Beslan and Peshawar school attacks makes precisely this argument—that outbidding led to these attacks (Biberman and Zahid 2019). In general, the outbidding argument is not necessarily about direct rivalry, such as intergroup violence, and is about more general competition over resources. It is also noteworthy that empirical support for outbidding leading to more violence is mixed (e.g., Findley and Young 2012). However, if rivalry is associated with outbidding among insurgent groups, it seems likely that schools would be exemplar targets.

On the other hand, beyond outbidding, rivalry suggests attacks on certain types of targets. Rivalry can manifest itself in attacks on civilians believed to be affiliated with the rival group. For example, interfiled rivalry between Republican and Loyalist groups in Northern Ireland led to attacks on pubs associated with the rival community. Rivalry between Sunni and Shi'ite groups sometimes leads to attacks on the mosques associated with the rival group, as discussed previously. Rivalry among the Colombian groups FARC and the National Liberation Army (ELN) resulted in killings of group leaders (Rochlin 2003, 124). FARC also attacked an ELN-controlled pipeline to encourage the ELN to share profits (Wilson 2003).

However, when thinking about types of targets, it is not clear that rivalry should lead to attacks on schools in particular. Rivals want to intimidate and possibly destroy each other, but schools are not the best types of targets for these goals. As rivals attack their enemies' group members and supporters, schools are often too broad or indiscriminate of a target to affect their enemy as intended. One exception would be in sectarian violence, interfield rivalry, groups could attack a school associated with the ethnic group of their rival. However, many schools are mixed and not only of one ethnic group. In addition, many rivalries are intrafield, among groups claiming to represent the same ethnic group or political goal, so a school could contain children of both rival insurgent groups. Overall, whether one thinks of outbidding or target types, it is not clear how rivalry should affect attacks on schools, so we pose two competing hypotheses:

> School H4a (outbidding): Intergroup rivalries are associated with a *higher* likelihood of educational institution targeting.
> School H4b (target type): Intergroup rivalries are *not* associated with educational institution targeting.

6.3.3. Interacting with the Public: Social Services, Ethnic Motivation, and Crime

In addition to insurgents' relationships with the state and other militant organizations, insurgents also have important relationships with the public. These relationships are likely to be important for understanding why some insurgents attack educational institutions. The main theoretical argument in Chapter 2 focused on social services and ethnic motivation as important indicators of insurgents' interactions with the public. The primary empirical tests of Chapter 4 found social service provision associated with terrorism use, and it would make sense that it is associated with the subset of attacks on educational institutions—and possibly to a greater degree than terrorism in general.

Insurgents who provide social services to communities have a special bond with recipients, providing club goods that can obligate community members to help the insurgents. This is why insurgents who provide such services are able to carry out more spectacular attacks, including suicide attacks (Berman and Laitin 2008; Flanigan 2006; Iannaccone and Berman 2006). It might seem counterintuitive that a militant group would both aid the community and attack its students. However, because groups providing social services are competing with the state, schools can be viewed as symbols of state service provision. This suggests schools would be especially valuable targets for groups providing social services. Furthermore, educational institutions could be likely to suffer attacks when insurgents view state education as promoting values or a national history that the insurgents dispute. Turkey's Kurdistan Workers Party (PKK) apparently had this idea in mind when they set fire to 20 empty schools in 2012 (Euronews 2012). The PKK also temporarily kidnapped teachers during the same period. The group had stated that Turkish educational institutions are an instrument to assimilate Kurdish children (Euronews 2012). Overall, groups that provide social services should be especially willing and able to attack educational institutions.

School H5: Social service provision is associated with a *higher* likelihood of educational institution targeting.

Chapter 4 demonstrated that insurgents with ethnic motivations are more likely to use terrorism compared to those with other types of motivations. As with interorganizational rivalry discussed previously, it is not clear how ethnic motivation should affect terrorism against educational institutions. One possibility is that groups claiming to represent a particular ethnic community often engage in outbidding (Bloom 2004, 2005), and this suggests more extreme attacks, including possibly attacking schools. This is consistent with work showing that ethnically motivated groups are especially likely to attack soft targets (Asal et al. 2009). Prominent cases of outbidding leading to suicide terrorism include ethnically motivated groups such as the Tamil Tigers, Hamas, Chechen separatists, and the PKK. However, Bloom also notes that outbidding only leads to suicide attacks when such extreme violence is acceptable to the community that supports the group. This is why some ethnically motivated groups, such as those in Northern Ireland, never used the tactic and why suicide bombing fell out of favor to other groups such as the PKK and the Tamil Tigers, according to Bloom (2005, 81–83). Across many ethnically motivated groups, however, outbidding has led to increasingly extreme violence.

An alternate possibility is that ethnic motivations might be orthogonal to attacks on schools. For many insurgent groups, a primary audience of their violence is the state. For these groups, attacking schools could be an appealing tactic. For other insurgents, such as groups with rivals, an important audience is this rival group. This could be the situation for ethnically motivated insurgents, who are often mobilized to perceive another ethnic community as an enemy. In the same way that interorganizational rivalry leads to terrorism against certain types of targets—for example, other militant groups and their perceived supporters—ethnic motivation is likely to be associated with sectarian terror. Sectarian violence can spiral because members of ethnic groups are identifiable in ways that supporters of a certain ideology are not (Chandra 2006) and therefore can be easy targets for groups claiming to represent the rival group. This has been shown qualitatively in many case studies (e.g., Bruce 1997; Nasr 2000). If this is the situation more broadly, then ethnic motivation should be associated with attacks on members and symbols of a competing ethnic community. However, these groups could become focused on sectarian struggle and not carry out terror on state-associated targets such as schools. This would suggest a lack of relationship between ethnic motivation and educational institution targeting. As with interorganizational rivalry, then, we pose two competing hypotheses:

School H6a (outbidding): Ethnic motivations are associated with a *higher* likelihood of educational institution targeting.

School H6b (target type): Ethnic motivations are *not* associated with educational institution targeting.

A final way that insurgent groups' relationships with the public might be associated with violence against educational institutions is through group involvement in crime. Groups involved in crime are likely to have a predatory relationship with the public. If insurgents are kidnapping, extorting, and robbing the public, they might have less restraint with regard to attacking especially sensitive targets, such as schools. Predatory insurgents are less likely to try to win public support because their relationship with average citizens is one coercive. It has also been shown that insurgent groups that depend on economic incentives for their members are more prone to lose control over their members—which often results in abuse against civilians (Weinstein 2006). These issues should apply more strongly to taboo violence, such as attacks on schools, than more typical terrorism, such as attacks on political actors.

In addition to less restraint for these kinds of insurgent organizations, and less control over their members, a relationship between crime and school targeting could be more direct. Groups engaged in crime might attack a school if extortion payments are not made or, more generally, to strike fear in the population in the service of securing submission to the group's authority. In addition, if a group is already engaging in kidnapping, a school represents a target full of easily kidnapped victims. This was the situation with Boko Haram's repeated attacks on schools in Nigeria. The group has an ideological opposition to aspects of schooling, including girls' education, but it also has used the attacks in a practical way, obtaining new recruits and cowing civilians into compliance with random demands. Overall, there are a number of causal mechanisms through which involvement in crime might be related to attacks on educational institutions. Because educational institutions in particular, as opposed to random public targets, can be useful in crime, the relationship here could be stronger than that for terrorism generally.

> **School H7**: Involvement in crime is associated with a *higher* likelihood of educational institution targeting.

Given the many hypotheses, we present Table 6.2 to show how the hypotheses of this chapter compare with those of Chapter 2. For several of the key explanatory factors in Chapter 2—government concessions and coercion, intergroup alliances, and social service provision—the relationship with the targeting of educational institutions is expected to be stronger than that between these factors and terrorism in general. However, for two other factors, rivalry and ethnic motivation, the expectations are unclear. These attributes could be associated with an increased likelihood of school targeting, to a greater degree than they are for terrorism generally. There are also reasons to expect that rivalry and ethnic motivations are not associated with attacks on educational institutions. The basic logic behind this latter argument is that

Table 6.2 School Hypotheses

Hypothesis	Key Independent Variable Concept	Expected Impact on School Terror
1	Concessions	Decreased likelihood
2	Coercion	Increased likelihood
3	Intergroup alliances	Increased likelihood
4a	Intergroup rivalry	Increased likelihood
4b (competing or null hypothesis)		No relationship
5	Social service	Increased likelihood
6a	Ethnic motivation	Increased likelihood
6b (competing or null hypothesis)		No relationship
7	Crime	Increased likelihood

rivalry and ethnic motivations suggest groups are focused on other militants more than the state, so they are not especially likely to attack state-symbolic soft targets such as schools. The following section empirically evaluates these hypotheses.

6.4. Empirical Analyses of Attacks Against Educational Facilities

This section presents a replication of the empirical models in Chapter 4, with the alternate and more specific dependent variable of terrorist attacks against educational institutions. Given that the data and models were described in Chapter 3 and 4, this information is not repeated here. The novel empirical aspect of this chapter, the dependent variable, derives from information in the GTD. To code *Terrorism against schools*, we examine all attacks associated with the groups in our data and code the variable "1" if the group carried out an attack on an education institution that year. The GTD has 22 target types, one of which is educational institution. As with *Terrorism* in the main models, we use a dichotomous measure because we are more interested in whether a group uses terrorism or not, as opposed to whether the group uses one additional attack—for example, four instead of three.

Regarding the distribution of the dependent variable, as noted previously, of the 140 groups in our data, 44 (31%) conducted an attack on an educational target at some point. However, only approximately 8% of the

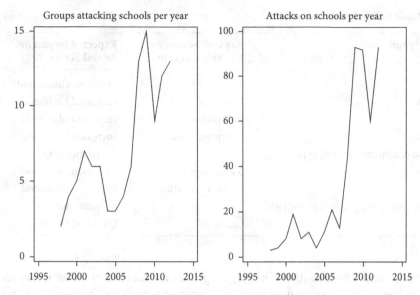

Figure 6.1 Insurgent groups' attacks on educational institutions over time.

group-year observations are coded for *Terrorism against schools*, indicating that the attacks are not very common. Although almost one-third of groups have carried out such an attack, they might do so only during one or a few years. Figure 6.1 shows there is a substantial temporal dynamic to these attacks because they increased substantially in approximately 2008. There had been no more than 20 attacks per year through 2007, and then in 2008 there were 43. In 2009, there were 93. Similarly, the number of groups using this tactic spiked at the same time, increasing from 6 groups in 2007 to 13 in 2008 and 15 in 2009. This is consistent with the broader trend beyond the civil conflict context. Terrorism against schools was relatively rare until approximately 2005 (Petkova et al. 2017). The fact that the number of groups using this tactic increased so quickly is suggestive of diffusion (Horowitz and Potter 2014).

6.4.1. Results

Table 6.3 replicates Table 4.8, replacing the dependent variable with *Terrorism against schools*. Model 1 contains only control variables, Model 2 includes only the key independent variables, and Model 3 is the full main

Table 6.3 Logit Models of Attacks on Educational Institutions by Insurgent Organizations

	Model 1	Model 2	Model 3
	Only Control Variables	Only Key Variables	Main Model
Stick		0.446	1.669**
		(0.300)	(0.532)
Alliances		0.155**	0.123**
		(0.055)	(0.058)
Rivalry		0.361	0.052
		(0.249)	(0.260)
Social services		1.732***	1.761***
		(0.464)	(0.464)
Ethnic motivation		0.012	−0.051
		(0.430)	(0.447)
Crime		0.276*	0.079
		(0.147)	(0.148)
Mixed carrot and stick	0.639*		1.922***
	(0.328)		(0.572)
Territory	0.706*		0.146
	(0.396)		(0.425)
Religious	1.396**		0.624
	(0.435)		(0.497)
Leftist	0.407		0.369
	(0.526)		(0.530)
Group size	0.313		0.401
	(0.266)		(0.268)
School terror history	0.009		0.009
	(0.025)		(0.025)
Conflict battle deaths	0.000*		0.000
	(0.000)		(0.000)
State terror	0.388*		0.282
	(0.230)		(0.239)
Democracy	0.192**		0.121
	(0.080)		(0.084)
GDP per capita	0.401*		0.390*
	(0.225)		(0.218)
Population (log)	0.105		0.140
	(0.130)		(0.128)
Constant	−12.910***	−5.014***	−13.749***
	(3.304)	(0.791)	(3.323)
N	1,240	1,240	1,240

Note: Standard errors in parentheses. Models include group random effects and year fixed effects. *$p < .10$, **$p < .05$, ***$p < .01$. All independent variables are lagged 1 year. This table replicates Table 4.8, in which *Terrorism* was the dependent variable.

model. The results appear somewhat similar to those of Table 4.8, but there are also a number of important differences. First, a note about variables included. *Carrot* is not included in the models because there are no years in which insurgent groups attacked a school and had been offered government concessions the previous year. There is a perfect correlation between the two variables, so this causes *Carrot* to drop from the models. Although we are not able to show a coefficient or otherwise indicate the magnitude of the relationship, it is a perfect negative relationship. Insurgent groups that receive concessions in one year *never* attack schools the subsequent year. This suggests strong support for Hypothesis 1. The one other variable that appeared in Table 4.8 but does not appear in Table 6.3 is *Single leader*, the indicator of insurgent groups led by one person. None of the groups with a single leader ever attack a school, so this variable cannot be included in the models.[4] All other variables from Table 4.8 are included.

Among the variables representing hypothesized relationships shown in the main model, *Stick*, *Alliance*, and *Social services* all have statistically significant and positive coefficients. This suggests support for Hypotheses 2, 3, and 5. This is also consistent with the findings in Chapter 4 for terrorism generally. Regarding substantive significance, Figure 6.2 uses marginal effects to compare the estimated impact of each independent variable. *Carrot* is not shown for reasons discussed previously. Of the variables representing hypothesized relationships, the .09 marginal effect on *Stick* suggests that if a group is subject to government coercion in the previous year, it is estimated to have a 9 percentage-point increase in the probability of attacking a school. Social service provision is associated with approximately the same effect, whereas the impact of alliances is only approximately a 1 percentage-point increase.

Interestingly, the coefficient on *Rivalry* is statistically insignificant. There is no evidence of a relationship between intergroup competition and terrorism against educational institutions. This raises questions about outbidding (Hypothesis 4a) and suggests support for the notion that rivalries might lead to attacks on rival groups and their perceived associates more than attacks on seemingly "neutral" or state-associated targets such as schools (Hypothesis 4b). It is also noteworthy that the coefficient on *Ethnic motivation* is statistically insignificant. This suggests there is no relationship between an insurgent group having an ethnic motivation and attacks on educational institutions. This casts doubt on the notion of ethnic outbidding leading to school attacks

---

[4] This is only 12 groups, less than 9% of the total.

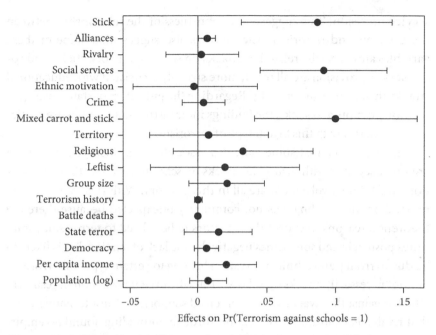

Figure 6.2 Substantive significance of variables in the main model. Average marginal effects are shown with 95% confidence intervals.

(Hypothesis 6a), and it is more supportive of the idea that ethnically motivated insurgent groups attack unequivocally sectarian targets, as opposed to schools (Hypothesis 6b). The coefficient on *Crime* is also statistically insignificant, suggesting that group involvement in crime such as drug trafficking or extortion is not related to subsequent terrorism by the group. School Hypothesis 7 is not supported. This is surprising because there are theoretical reasons to expect crime would be related to school attacks. In addition, the relationship between crime and terrorism discussed in Chapter 4 was one of the most robust relationships. Apparently, school targeting is indeed a unique type of terrorism, with its own particular drivers. This is discussed further in the next section.

6.4.2. Discussion

The previously discussed findings are interesting for their similarities and contrasts with the findings of Chapter 4, that examine terrorism in general. Regarding similarities, several factors are shown to be related to both all terrorism and attacks on schools: carrots, sticks, intergroup alliances, and social service provision. This suggests the robustness of the relationship between these factors and terrorism. The findings also suggest that some of these variables are especially related to attacks on schools. Carrots, sticks, and social service provision are all much more strongly associated with educational attacks than terrorism generally. Regarding the indicators of state–insurgent relations, carrots and sticks, the findings indicate that state actions are particularly powerful with this brutal type of terrorism.

It is remarkable that some variables associated with insurgent terrorism are not associated with insurgent attacks on schools. Rivalry, ethnic motivation, and crime involvement are all in this category. With rivalry and ethnic motivation, this finding was not completely unexpected because there are theoretical reasons to expect these factors to be related to terrorism, sometimes postively and sometimes negatively. The lack of relationship is likely to be due to rivalry and ethnic motivation leading to certain types of terrorism, but not necessarily attacks on schools. Inter-insurgent rivalry often results in attacks against the rival group or perceived supporters. The alternative story, that rivalry leads to attacks on schools through outbidding, found no empirical support, which raises questions about outbidding theory. It is especially noteworthy that a recent case study argues that outbidding is precisely what explains several prominent attacks on schools (Biberman and Zahid 2019). Other research has suggested nuance about the conditions in which outbidding leads to more terrorism (Brym and Araj 2008; Findley and Young 2012), and our findings here suggest a need for additional research on the often-discussed notion of outbidding.

On outbidding, it should be acknowledged that Bloom (2005) originally suggested nuance with the theory—that it is not competition alone that leads to unusual violence. She argued that an important conditioning factor is societal acceptance of violence. More recent work suggests other types of nuance—for example, Biberman and Zahid's (2019) contention that outbidding should lead to extreme violence when there are internal group strains or external pressure from the state. Overall, the complex dynamics theorized to be behind outbidding could be viewed as consistent with our finding of a lack of a general relationship between group rivalry and attacks on schools. In certain conditions, intergroup competition does seem to lead to extreme violence. But this does not appear to be the case generally for insurgent groups in recent decades.

Regarding ethnic motivation, these types of groups might be especially likely to attack symbols of a rival ethnic group, but this does not often translate into attacks on schools. Insurgents with ethnic motivations are per-haps more likely to target the government that might oppress them, as well as a rival ethnic group. Schools are not necessarily valuable targets to ethnically motivated insurgents. This, too, raises questions about outbidding because prominent cases of outbidding are ethnically motivated groups, such as militant organizations claiming to represent Tamils, Palestinians, or Kurds. Ethnic motivation alone seems to be insufficient to explain outbidding.

Finally, the lack of a relationship between the crime measure and attacks on schools is striking given the robustness of the crime relationship discussed in Chapter 4. One explanation could be that school terrorism simply has unique explanatory factors relative to terrorism in general. A more nuanced story is that insurgent groups that engage in crime are sometimes more on the "criminal" end of the spectrum than purely politically motivated. As a result, they might prefer to keep profiting off crime and not draw ex-cessive government attention by conducting severe attacks such as those against schools. They would use terrorism occasionally to punish a rival or demonstrate their continued relevance, but they do not go too far because they do not want to reduce their illicit gains. This is consistent with the no-tion of "transformation" into a criminal group or "hybridization" between crime and militancy (Dishman 2001; Makarenko 2004; Shelley 2014). Two examples of groups engaged in crime and some terrorism, but with almost no attacks on schools, are the People's United Liberation Front (PULF) of India and the Real Irish Republican Army (RIRA).[5] Both are substantially involved in extortion and other crimes (Cusack 2010; Hourigan et al. 2018; South Asia Terrorism Portal 2012).[6] The notion of engaging in crime and ter-rorism, but not extreme terrorism such as school attacks, might be an indi-cator that these groups are more on the "criminal" end of the spectrum than "terrorist." This is consistent with some analysts' views of these two groups (e.g., Jupp and Garrod 2019). Looking at attacks on schools, then, is helpful for sorting out what might be described as "real terrorists" from groups that use terrorism less frequently or in a less "pure" sense—for example, attacks against highly vulnerable civilians such as students.

[5] PULF has no attacks on educational institutions in our data, but RIRA carried out one such at-tack. However, the one RIRA attack on a school was in 2001 (Laville 2001), early in its career, before it had transitioned to a mostly criminal group. This is consistent with the idea of attacks on schools being associated with more purely "political" groups and less so with mostly "criminal" groups.

[6] *Forbes* noted in 2018 that RIRA, a relatively small organization, was one of the richest militant groups in the world (Zehorai 2018).

The idea that insurgents engaged in crime would be careful about provoking the state suggests an important relationship between insurgents and government. However, it adds nuance to the causal relationship theorized in Chapter 3 about criminal insurgents that prey on the population and therefore are unlikely to show restraint toward civilians. This mechanism may be generally true (as the results in Chapter 4 suggest), but the results regarding school attacks suggest there are limits on this lack of restraint. The insurgent–civilian relationship regarding crime seems to be conditioned by concerns about the insurgent–state relationship. If insurgents abuse civilians too much—if they attack the most vulnerable populations—the state must respond. Many insurgent groups substantially involved in crime probably seek to avoid such a state response because it would interfere with their business activities.

6.5. Conclusion

This chapter expanded upon the main findings of the book, presented in Chapter 4, by exploring how insurgent group embeddedness might be related to a specific and important type of terrorism—attacks on educational institutions. These types of targets are important to study because they represent an extreme type of terrorism: attacks against soft targets that are not directly connected to armed conflict. The chapter theorized about how indicators of insurgent embeddedness should be related to attacks on schools. Empirical tests found four interesting results.

First, some factors related to terrorism are also related to school attacks. State coercion, intergroup alliances, and social service provision are also associated with an increased likelihood of insurgent group terrorist attacks on educational institutions. A fourth factor, government concessions, could not be included in statistical tests because there are no instances of government concessions and attacks on schools during the same group-year. Despite not being able to empirically test for this relationship, however, this perfect correlation suggests support for the idea that government concessions reduce the likelihood of attacks on schools. Overall, the consistency of the relationships of state coercion, state concessions, alliances, and social service provision with both terrorism and the extreme terrorism type of school targeting suggest these factors are highly important for understanding insurgent terrorism.

Second, in terms of substantive significance, some of these factors are much more strongly related to school terrorism than terrorism in general. This is consistent with theorized causal mechanisms. For example, if concessions lead groups to abstain from terrorism, it should make them especially likely to abstain from extreme terrorism—unusually shocking terrorism such as school attacks. In addition, social service provision was shown to have a much stronger relationship with school terrorism than terrorism generally. This could be because groups providing social services view state social services such as schools as direct competition with their own quasi-governance, so they are especially likely to attack these targets. To our knowledge, no one has studied the relationship between social service provision and certain types of terrorist attacks, such as violence against state services. This is worthy of future research.

Third, several insurgent group attributes robustly related to terrorism in Chapter 4 were found *not* to be related to terrorism against schools. On the one hand, it raises questions about robustness. On the other hand, it points to the fairly unique nature of terrorism against schools, and it is suggestive of the logic behind the relationship between these factors and terrorism and helps narrow down causal mechanisms. For two variables, intergroup rivalry and ethnic motivation, it was hypothesized that they should be associated with terrorism against rival organizations and the communities that their rivals seek to represent. As a result, it makes sense that rivalry and ethnic motivation are not related to school terrorism. Schools are unlikely to contain members of rival groups, and they are also unlikely to have pupils of only one ethnicity; they make poor targets for group rivalry or ethnic-based violence. Interestingly, one reason rivalry and ethnic motivation could be related to school terrorism would be outbidding—group competition manifesting itself in the relatively extreme school attacks. The lack of association between these factors and terrorism against educational institutions suggests more nuanced research on outbidding is needed.

A fourth and final interesting finding is the lack of relationship between crime and terrorism against schools. It was hypothesized that groups that engage in crime have a predatory relationship with the public and therefore will be likely to also attack schools. However, it was noted that many insurgent groups involved in crime are "hybrid" militant–criminal groups that are more focused on illicit gains than affecting politics, so they might avoid extreme terrorism to not incur the full wrath of the state. The notion that different degrees of involvement in crime might lead to distinct insurgent group consequences has not been researched sufficiently, so this seems like another important path for future research.

Overall, examining terrorism against educational institutions allows us to further think about the relationship between insurgent embeddedness and civilian targeting. Insurgents' relationships with the state, other insurgents, and the public have important implications not just for terrorism in general but also for more extreme types of terrorism. Examining an example of such extreme terrorism, school targeting, gave us the opportunity to test the limits of the arguments and the causal mechanisms behind them. Insurgent groups' relationships are crucial for understanding terrorism by insurgents, and different types of terrorism are affected in distinct ways.

References

Abrahms, Max. "Why Terrorism Does Not Work." *International Security* 31, no. 2 (2006): 42–78.

Akcinaroglu, Seden. "Rebel Interdependencies and Civil War Outcomes." *Journal of Conflict Resolution* 56, no. 5 (2012): 879–903.

Asal, Victor, and R. Karl Rethemeyer. "The Nature of the Beast: Organizational Structures and the Lethality of Terrorist Attacks." *Journal of Politics* 70, no. 2 (2008): 437–49.

Asal, Victor H., R. Karl Rethemeyer, Ian Anderson, Allyson Stein, Jeffrey Rizzo, and Matthew Rozea. "The Softest of Targets: A Study on Terrorist Target Selection." *Journal of Applied Security Research* 4, no. 3 (2009): 258–78.

Berman, Eli, and David D. Laitin. "Religion, Terrorism and Public Goods: Testing the Club Model." *Journal of Public Economics* 92, nos. 10–11 (2008): 1942–67.

Biberman, Yelena, and Farhan Zahid. "Why Terrorists Target Children: Outbidding, Desperation, and Extremism in the Peshawar and Beslan School Massacres." *Terrorism and Political Violence* 31, no. 2 (2019): 169–84.

Bloom, Mia M. "Palestinian Suicide Bombing: Public Support, Market Share, and Outbidding." *Political Science Quarterly* 119, no. 1 (2004): 61–88.

Bloom, Mia. *Dying to Kill: The Allure of Suicide Terror*. New York, NY: Columbia University Press, 2005.

Bradford, Emma, and Margaret A. Wilson. "When Terrorists Target Schools: An Exploratory Analysis of Attacks on Educational Institutions." *Journal of Police and Criminal Psychology* 2, no. 28 (2013): 127–38.

Brandt, Patrick T., and Todd Sandler. "What Do Transnational Terrorists Target? Has It Changed? Are We Safer?" *Journal of Conflict Resolution* 54, no. 2 (2010): 214–36.

Bruce, Steve. "Victim Selection in Ethnic Conflict: Motives and Attitudes in Irish Republicanism." *Terrorism and Political Violence* 9, no. 1 (1997): 56–71.

Brym, Robert J., and Bader Araj. "Palestinian Suicide Bombing Revisited: A Critique of the Outbidding Thesis." *Political Science Quarterly* 123, no. 3 (2008): 485–500.

Chandra, Kanchan. "What Is Ethnic Identity and Does It Matter?" *Annual Review of Political Science* 9 (2006): 397–424.

Cusack, Jim. "RIRA Expanding its Crime Empire." *Irish Independent.* August 15, 2010. https://www.independent.ie/irish-news/rira-expanding-its-crime-empire-26672174.html

Dishman, Chris. "Terrorism, Crime, and Transformation." *Studies in Conflict and Terrorism* 24, no. 1 (2001): 43–58.

Euronews. "Alleged PKK Members Attack Schools in Turkey." October 23, 2012.

Eyerman, Joe. "Terrorism and Democratic States: Soft Targets or Accessible Systems." *International Interactions* 24, no. 2 (1998): 151–70.

Fahey, Susan, and Victor Asal. "Lowest of the Low: Why Some Countries Suffer Terrorist Attacks Against Schools." *Dynamics of Asymmetric Conflict* 13, no. 2 (2020): 101–24.

Findley, Michael G., and Joseph K. Young. "More Combatant Groups, More Terror? Empirical Tests of an Outbidding Logic." *Terrorism and Political Violence* 24, no. 5 (2012): 706–21.

Flanigan, Shawn Teresa. "Charity as Resistance: Connections Between Charity, Contentious Politics, and Terror." *Studies in Conflict & Terrorism* 29, no. 7 (2006): 641–55.

Fortna, Virginia Page. "Do Terrorists Win? Rebels' Use of Terrorism and Civil War Outcomes." *International Organization* 69, no. 3 (2015): 519–56.

Fujii, Lee Ann. "The Puzzle of Extra-Lethal Violence." *Perspectives on Politics* 11, no. 2 (2013): 410–26.

Horowitz, Michael C., and Philip BK Potter. "Allying to Kill: Terrorist Intergroup Cooperation and the Consequences for Lethality." *Journal of Conflict Resolution* 58, no. 2 (2014): 199–225.

Hourigan, Niamh, John F. Morrison, James Windle, and Andrew Silke. "Crime in Ireland North and South: Feuding Gangs and Profiteering Paramilitaries." *Trends in Organized Crime* 21, no. 2 (2018): 126–46.

Iannaccone, Laurence R., and Eli Berman. "Religious Extremism: The Good, the Bad, and the Deadly." *Public Choice* 128, nos. 1–2 (2006): 109–29.

International Peace Conference. "Convention (IV) Respecting the Laws and Customs of War on Land and Its Annex: Regulations Concerning the Laws and Customs of War on Land." The Hague, the Netherlands, October 18, 1907. https://ihl-databases.icrc.org/ihl/INTRO/195.

Jupp, John, and Matthew Garrod. "Legacies of the Troubles: The Links Between Organized Crime and Terrorism in Northern Ireland." *Studies in Conflict & Terrorism* (2019): 1–40.

Kalyvas, Stathis N. *The Logic of Violence in Civil War.* Cambridge University Press, 2006.

Kydd, Andrew H., and Barbara F. Walter. "The Strategies of Terrorism." *International security* 31.1 (2006): 49–80.

Laville, Sandra. "Ta Cadet in Bomb Blast Loses Sight." *The Telegraph.* February 24, 2001. https://www.telegraph.co.uk/news/uknews/1323948/TA-cadet-in-bomb-blast-loses-sight.html.

Makarenko, Tamara. "The Crime–Terror Continuum: Tracing the Interplay Between Transnational Organised Crime and Terrorism." *Global Crime* 6, no. 1 (2004): 129–45.

Mashal, Mujib, and Salman Masood. "Bomb Strikes Pakistan Mosque Frequented by Afghan Taliban Chief." *The New York Times.* August 16, 2019. https://www.nytimes.com/2019/08/16/world/asia/pakistan-taliban-mosque-bomb.html.

Moghadam, Assaf, Ronin Berger, and Polina Beliakova. "Say Terrorist, Think Insurgent: Labeling and Analyzing Contemporary Terrorist Actors." *Perspectives on Terrorism* 8, no. 5 (2014): 2–17.

Montaño, John. Internado Destruido por FARC Dejó en el Aire a Niños de Tres Veredas." *El Tiempo* (Colombia). February 4, 2013. https://www.eltiempo.com/archivo/docume nto/CMS-12578596.

Myers, Steven Lee. "Truck Bombing at Russian Military Hospital Kills 35; Officials Blame Chechen Separatists." *The New York Times*. August 2, 2003. https://www.nytimes.com/ 2003/08/02/world/truck-bombing-russian-military-hospital-kills-35-officials-blame-chechen.html.

Nasr, Vali R. "International Politics, Domestic Imperatives, and Identity Mobilization: Sectarianism in Pakistan, 1979–1998." *Comparative Politics* (2000): 171–90.

Petkova, Elsiveta P., Stephanie Martinez, Jeffrey Schlegelmilch, and Irwin Redlener. "Schools and Terrorism: Global Trends, Impacts, and Lessons for Resilience." *Studies in Conflict & Terrorism* 40, no. 8 (2017): 701–11.

Polo, Sara M. T. "The Quality of Terrorist Violence: Explaining the Logic of Terrorist Target Choice." *Journal of Peace Research* 57, no. 2 (2020): 235–50.

Polo, Sara M. T., and Kristian Skrede Gleditsch. "Twisting Arms and Sending Messages: Terrorist Tactics in Civil War." *Journal of Peace Research* 53, no. 6 (2016): 815–29.

Reuters. "Islamic State Says It Carried Out Attack at Shi'ite Mosque in Central Afghanistan." July 6, 2019. https://www.reuters.com/article/us-afghanistan-blast/ islamic-state-says-it-carried-out-attack-at-shiite-mosque-in-central-afghanistan-idUSKCN1U109O.

Rochlin, James F. *Vanguard Revolutionaries in Latin America: Peru, Colombia, Mexico.* Boulder, CO: Lynne Rienner, 2003.

Santifort, Charlinda, Todd Sandler, and Patrick T. Brandt. "Terrorist Attack and Target Diversity: Changepoints and Their Drivers." *Journal of Peace Research* 50, no. 1 (2013): 75–90.

Schmid, Alex P., and Albert J. Jongman. *Political Terrorism: A New Guide to Actors, Authors, Concepts, Data Bases, Theories, & Literature.* NJ: Transaction, 2005.

Shelley, Louise I. *Dirty Entanglements: Corruption, Crime, and Terrorism.* Cambridge, UK: Cambridge University Press, 2014.

South Asia Terrorism Portal. "Incidents and Statements Involving PULF: 1998–2012." 2012. https://www.satp.org/satporgtp/countries/india/states/manipur/terrorist_outf its/PULF_tl.htm.

Thomas, Jakana. "Rewarding Bad Behavior: How Governments Respond to Terrorism in Civil War." *American Journal of Political Science* 58, no. 4 (2014): 804–18.

Tokdemir, Efe, and Seden Akcinaroglu. "Reputation of Terror Groups Dataset: Measuring Popularity of Terror Groups." *Journal of Peace Research* 53, no. 2 (2016): 268–77.

UN General Assembly. *Universal Declaration of Human Rights* (217 [III] A). Paris, 1948. https://www.un.org/sites/un2.un.org/files/udhr.pdf

Weinstein, Jeremy M. *Inside Rebellion: The Politics of Insurgent Violence.* Cambridge, UK: Cambridge University Press, 2006.

Wilson, Scott. "U.S. Moves Closer to Colombia's War: Involvement of Special Forces Could Trigger New Wave of Guerrilla Violence." *The Washington Post.* February 7, 2003.

Zehorai, Itai. "The Richest Terrorist Organizations in the World." *Forbes.* January 24, 2018. https://www.forbes.com/sites/forbesinternational/2018/01/24/the-richest-ter ror-organizations-in-the-world.

7

Why Do Some Insurgent Groups Attack Journalists?

7.1. Introduction

On October 12, 2015, the Taliban announced that it would no longer recognize two Afghan television channels as "media outlets," and that it was designating them as "military objectives," because it said they had reported falsely about Taliban attacks (Committee to Protect Journalists 2015).[1] Three months later in Kabul, a suicide bomber detonated a powerful explosion near a bus carrying nearly 30 members of Kaboora Production, an affiliate of one of the Taliban-blacklisted stations. The explosion killed at least 7 people and injured 25 others. The Taliban claimed responsibility for the attack and warned of more attacks if news agencies did not change their policies (Jolly and Sukhanyar 2016).[2] The Taliban is far from alone in its attacks on journalists. Other insurgents, such as al-Shabaab and Colombia's National Liberation Army (ELN) have kidnapped, tortured, or killed journalists or damaged buildings in which journalists work. Why do insurgents sometimes attack the news media? In other words, why do they "kill the messenger"?

This chapter provides another extension of the theory and empirical tests, exploring potential links between insurgent embeddedness and attacks against journalists. Unlike attacks on schools discussed in Chapter 6, this is not an "extreme test" of the model because the news media are probably not as much of an extreme or taboo target for insurgents as are children. However, journalists are an important and interesting type of target for a number of reasons. A unique logic is likely to explain why insurgents occasionally attack

---

[1] The full Taliban statement is available at https://cpj.org/wp-content/uploads/2016/03/Taliban-threat3.pdf.

[2] Interestingly, the Taliban was careful in its language about attacking journalists, implying that "authentic" journalists would not be targeted. This is consistent with how they treated a 2016 attack, which killed two journalists—one American and one Afghan. The group stated that it had not realized the victims were journalists (Little 2017). This points to the relatively unique status that journalists have among potential targets for insurgents.

Insurgent Terrorism. Victor Asal, Brian J. Phillips, and R. Karl Rethemeyer, Oxford University Press. © University of Maryland National Consortium for the Study of Terrorism and Responses to Terrorism (START) 2022.
DOI: 10.1093/oso/9780197607015.003.0007

members of the news media. Learning if and how embeddedness characteristics relate to attacks against journalists helps us better understand both
the dynamics of insurgent embeddedness and violence against journalists
generally.

Violence against journalists is crucial to understand, but it is surprisingly understudied in the conflict literature, with few exceptions (Carey and
Gohdes 2017, 2021; Holland Rios 2017; Lopez 2016). One study examines
two terrorist groups and finds that they attack journalists for collaborating
with the enemy and in retaliation for negative coverage (Lopez 2016).
However, it is unclear if these findings are generalizable. In media studies,
scholars have analyzed questions such as why some geographic areas are
more dangerous than others for journalists (Brambila 2017) and to what
extent violence against the press is "criminal governance" (Charles 2020).
Journalists play an essential role in governance because they help citizens understand what is happening in their country and the world. The information
provided by journalists can be an important check on the power of government and other elite actors. Furthermore, attacks on journalists can be an
important indicator of systemic human rights concerns (Carey and Gohdes
2017). According to the Committee to Protect Journalists (2020), 1,376
journalists were killed between 1992 and mid-2020. Consistent with our
data, killings increased substantially since the late 1990s—from 24 in 1998
to 74 in 2012 (Committee to Protect Journalists 2020). Often the perpetrator
of violence against journalists is unknown. However, when the perpetrator
is identified, militant groups are often at fault: militant groups have killed
more journalists (460) than other groups, such as criminal organizations
(140) or government officials (222) (Committee to Protect Journalists 2020).
With militant groups, the government, and the military responsible for many
journalists' killings, it is clear that the civil conflict environment is dangerous
for journalists. Yet during many years, according to our data, militant groups
do *not* kill journalists.

The next section discusses the significance of attacks on journalists in civil
conflict. Section 7.3 considers the logic behind why insurgents attack news
media institutions and presents related hypotheses. Section 7.4 discusses empirical tests and their results. The final section concludes with implications of
the findings and a discussion of differences between these results and those
of previous chapters. Overall, the results of this chapter should be viewed
as somewhat exploratory because this is the first quantitative analysis of violence against journalists by insurgent groups. But the results provide an

opportunity to consider how the model might apply to this unique target type and, it is hoped, spur additional research on this outcome of interest.

7.2. The News Media as a Target Type

Journalists are under threat throughout the world, both in relatively peaceful developed countries (Löfgren Nilsson and Örnebring 2016; Obermaier, Hofbauer, and Reinemann 2018) and in countries experiencing substantial violence generally (Benítez 2017; Charles 2020). Since 1992, at least 20 journalists have been killed per year, with death tolls reaching the 70s in multiple years starting in 2007. Members of news organizations have often been attacked by governments—for example, when Saudi agents killed Jama Ahmad Khashoggi in the Saudi consulate in Istanbul in 2018 (Harris, Miller, and Dawsey 2018).[3] Drug cartels have also assaulted journalists in countries such as Mexico (Holland and Rios 2017; Relly and González de Bustamante 2014). More relevant to this book, insurgent organizations frequently attack journalists as well. In our data, 29 groups, approximately 21% of the 140 insurgent organizations in the data set, attack news media targets at some point.

Attacks on the news media may seem like any other kind of attack on civilians perpetrated by insurgent groups. However, they differ in important ways from attacks on civilians in general and from the subject of Chapter 6, schools. The news media are certainly noncombatants and prohibited targets, but attacks on the media are perhaps not as "extreme" as attacks on schools because attacking children is viewed as especially taboo. Attacks on journalists are not as extreme because the victims are generally adults.[4] The key difference between assaults on the news media and other kinds of terrorism relates to the role the news media play in society. Their role as communicators and potential checks on the government suggests

---

[3] Distinct logics motivate different actors' violence toward journalists, and the methods of violence vary as well. Salazar (2019) argues that government officials are more likely to beat journalists, whereas criminals are more likely to murder them. Different motivations behind these and other actors' violence toward the media are discussed more later.

[4] The 1907 Hague Convention mentions protection for journalists, noting that "individuals who follow an army without directly belonging to it, such as newspaper correspondents and reporters" are entitled to be treated as prisoners of war (International Peace Conference 1907, Article 13). The Protocol I amendment in 1977 notes that journalists should be considered as civilians, meaning they deserve the same protections as all other noncombatants (International Committee of the Red Cross 1977). It is important to note that in contrast with schools as discussed in Chapter 6, there is no explicit prohibition on attacking journalists other than indicating that journalists should be treated as any other civilians.

that violence against them is often fundamentally about communication—attacks are intended to either *reduce* or *shape* public information about the perpetrator insurgent organization.

Sometimes insurgents attack news media targets to simply silence them. This could be because the journalists are reporting about the insurgents in a way they do not prefer or because the journalists are viewed as pro-government (even if they do not mention the insurgents). For example, in Somalia, al Shabaab claimed responsibility for the 2010 killing of a prominent journalist; his colleagues believe he was attacked simply because he worked for government-run media (BBC 2016), which was critical of groups such as al Shabaab (BBC 2010). More recently, gunmen suspected to be members of al Shabaab killed a journalist (and his wife) who worked for a pro-government radio station (Al Jazeera 2015).[5] Overall, al Shabaab has been relatively prolific in its attacks on journalists, with 24 attacks recorded in our data.

In Indonesia in 2001, Free Aceh Movement members abducted three technicians from the state television network TVRI after accusing the station of pro-government coverage (Committee to Protect Journalists 2002). Journalists at state-run media organizations, or journalists at outlets viewed as friendly to the government, are often threated by insurgents.

Journalists also are attacked as insurgents try to shape news about them (Lopez 2016). In 2012 in Mali, members of the Azawad National Liberation Movement attacked the offices of a radio station, forcing it off the air for 3 days. They kidnapped and beat journalist Malick Ali Maiga, telling him to provide more positive coverage of the group in the future (Committee to Protect Journalists 2012).[6] Another example is the 1999 bombing of a newspaper office by the Revolutionary Armed Forces of Colombia (FARC), which was apparently upset at the newspaper's recent reporting (*El Tiempo* 1999). Later in Colombia, when unknown actors dynamited a radio tower, it was noted that the radio station had been airing government messages offering rewards for information about ELN (FLIP 2005). A more complex situation occurred in Pakistan in 2008, when a prominent columnist was murdered because the Baloch Liberation Army found the headline on his column insulting (Abbas 2013; Reporters Without Borders 2008).[7] Violence to affect

[5] This is Global Terrorism Database (GTD) incident 201505040044, described at https://www.start.umd.edu/gtd/search/IncidentSummary.aspx?gtdid=201505040044.

[6] Maiga was kidnapped again later in the same year for the same reason (Reporters Without Borders n.d.).

[7] This group is accused of killing multiple journalists throughout the years (e.g., Ahmed and Burke 2015).

news coverage is also consistent with the Taliban example presented at the beginning of the chapter, as the group told the journalists to report more accurately about it (Jolly and Sukhanyar 2016). This effort to shape media coverage is somewhat similar to the motives behind criminal group attacks on journalists. Criminals do not seem to mind journalists generally but attack them when their reporting is unfavorable or draws unwanted attention to them (Holland and Rios 2017). Controlling media coverage, then, is important for both insurgents and criminals.

It is noteworthy that although insurgents attack journalists to decrease or change coverage of themselves, they usually do not conduct such attacks to *increase* their group's coverage. This is interesting because terrorism is often thought of as an attempt by violent non-state actors to draw attention—to "dramatize a cause" (Crenshaw 1981, 389). Perhaps this is more common with terrorist groups and less common with insurgent organizations because the former are trying to establish a reputation, make a name, whereas insurgents are usually larger and have an established reputation.

In addition to the strategic communicative nature of attacks on the news media, these assaults occur for a variety of other reasons. Sometimes attacks on journalists occur in the form of kidnapping for ransom. One German journalist for *Der Spiegel* had the misfortune of being kidnapped twice during the same month. The first time was by Abu Sayyaf and the second was apparently by a related splinter group. He reportedly paid ransom both times (Kirk 2000). Insurgents might also suspect journalists are spies, especially if they are operating in an area the group considers its own territory. This happened with two Dutch journalists kidnapped by the ELN in Colombia in 2017 (BBC 2017).[8]

Overall, journalists play key roles in society and are important "target types" for insurgents because of their centrality to public communication. Insurgents sometimes attack journalists to silence them and in other instances to coerce them to report the news differently. There are also additional reasons, such as kidnapping for ransom, but much of the violence against journalists by insurgent groups seems to have a logic of communication behind it. The next section considers factors that might explain why some insurgent groups are more likely than others to attack journalists.

---

[8] This is GTD incident 201706170005, described at https://www.start.umd.edu/gtd/search/IncidentSummary.aspx?gtdid=201706170005.

7.3. Insurgent Embeddedness and News Media Targeting

Although the literature discusses violence against journalists, it remains unclear when we might expect violence against journalists to be more common by insurgent organizations. Much of the work on attacks against members of the news media has focused on the criminal context (Holland and Ríos 2017), and criminal violence is important in different ways from insurgent violence (Kalyvas 2015; Phillips 2015). One study examines the "terrorist organizations" ETA and ISIS and finds that these groups attack journalists for collaborating with the enemy and in response to negative portrayals (Lopez 2016). We are unaware of any studies seeking to explain *insurgent* violence against journalists in general. As a result, this section considers how the embeddedness theory of insurgent terrorism can be applied to this particular type of violence—insurgent attacks against journalists.

As with the previous two chapters, certain modifications are necessary to apply the embeddedness theory to attacks on the news media. Attacks on the news media are a quite specific type of violence, and therefore not all hypotheses from Chapter 2 might apply. Attacks on journalists are related to public communication, and any explanation of violence against the media must take this into consideration. The following sections consider how the likelihood of insurgent attacks on journalists is affected by the seven factors hypothesized to affect insurgent terrorism generally: government concessions, government coercion, intergroup alliances, intergroup rivalry, social service provision, ethnic motivation, and crime. Table 7.1 summarizes the hypotheses for the reader's convenience.

Table 7.1 News Media Hypotheses

Hypothesis	Key Independent Variable Concept	Expected Impact on Journalist Attacks
1	Concessions	Decreased likelihood
2	Coercion	Increased likelihood
3	Intergroup alliances	Increased likelihood
4	Intergroup rivalry	Increased likelihood
5	Social service provision	No relationship
6	Ethnic motivation	No relationship
7	Crime	Increased likelihood

7.3.1. Insurgents, the Targeted State, and Attacks on Journalists

Chapter 2 argued that interactions with the state—specifically carrots and sticks from the state—are likely to affect insurgent terrorism. Regarding attacks on journalists, the impact is likely to be similar. Concessions to insurgent groups are associated with a lower likelihood of terrorism generally. When the government has offered a "carrot" to insurgents, they should act in a way that does not discourage the state from providing similar benefits in the future. Given that journalists are generally "prohibited" targets in conflict, an attack on the news media after a show of good will from the state should be especially unlikely.

News media H1: Concessions to insurgent groups are associated with a *lower* likelihood of news media targeting.

Coercive actions toward insurgents are associated with a higher chance of subsequent terrorism by the insurgents, and the likelihood of attacks on the news media should probably be higher as well. Sometimes states are passive toward insurgents or offer them concessions (discussed previously), but in other instances they actively crack down on and pursue the insurgents. In this latter situation, insurgents are likely to respond—to signal to the government that there are costs to coercion. Attacks on the news media are an especially visible signal and therefore likely to be noticed by the government. As a result, it seems probable that insurgents would be more likely to attack journalists after government coercion as opposed to in other situations.

News media H2: Coercion toward insurgent groups is associated with a *higher* likelihood of news media targeting.

7.3.2. Intergroup Alliances, Rivalry, and Attacks on Journalists

For at least two reasons, alliances among insurgents are likely to be related to attacks on journalists. First, alliances can be an indicator of strength—of capability aggregation. Groups with allies tend to last longer (Phillips 2015) and carry out more attacks generally (Asal and Rethemeyer 2008; Horowitz and

Potter 2012). This increased capacity would be helpful for attacking news media targets. Although journalists are "soft" targets and generally unarmed, it seems likely that stronger groups are more likely to attack them. Weaker groups might attack random civilians or government employees to draw attention—through the resulting media coverage—to their attacks. But stronger insurgent groups feel confident enough to try to coerce journalists into providing the coverage they want. It is a substantial show of force to demand that the news media change its coverage to a group's liking, and strong and well-connected insurgent groups such as the FARC and the Taliban have done just this.

A second reason alliances should be associated with attacks on the press is that these attacks are a relatively unusual tactic and therefore likely to be learned through networks. It has been demonstrated that insurgent groups learn from each other, especially from those that are their allies (Cunningham, Dahl, and Frugé 2017;Horowitz 2010; Linebarger 2016). The learning could occur in the form of actual technical learning—that is, understanding how precisely attacks on journalists are most effectively carried out and why they are efficacious. Learning can also occur in the sense of normalization, in which it had been taboo to attack journalists but after allies have done it, a group decides to do it also. Either way, groups with more allies are more likely to have a partner who has already attacked a news media target and therefore are likely to subsequently do so themselves.

> **News media H3**: Intergroup alliances are associated with a *higher* likelihood of news media targeting.

Rivalries among insurgent groups are associated with an increased likelihood of civilian targeting generally, and it seems reasonable that they are also associated with attacks on journalists. There are multiple causal mechanisms through which competition can lead to terrorism. The most commonly debated is outbidding (Bloom 2004, 2005), in which groups use more extreme types of violence to draw attention and increase their popularity. This could apply to attacks on the news media, where groups in competition with others attack an especially visible and somewhat sensitive target—journalists. Note that journalists are not as extreme or taboo of a target as children, so if there is a relationship between rivalry and journalist attacks, it is probably not as substantial as the relationship between rivalry and attacks on schools. However, it seems that rivalry could lead to attacks on journalists through outbidding. Another possible mechanism is that rivalry causes groups to attack potential supporters of their rivals. An example of this is when Hamas gunmen set fire to a radio station and threatened others (*The Star* 2007). This kind of dynamic could explain some attacks on media associations seen to be favorable toward one group or another, but it might not be sufficient to explain a more general relationship between competition and attacks on journalists. If the outbidding mechanism is correct, however, we should expect rivalry to lead to attacks on the news media.

News media H4: Inter-group rivalries are associated with a *higher* likelihood of news media targeting.

7.3.3. Interacting with the Public: Social Services, Ethnic Motivation, and Crime

In addition to insurgents' interactions with the state and with other insurgents, their relationships with the broader public could also have implications for attacks on journalists. Social service provision, ethnic motivation, and involvement in crime are all related to terrorism generally, and this section considers how they might be related to assaults on the news media. All of these measures indicate a certain closeness with or dependence on the public, or at least a portion of the public in the case of ethnic motivation. As a result, we might expect any of these attributes to be related to journalist targeting because groups that depend on the public care more about their reputation, and they might coerce the news media to build or protect their public image. So generally we might expect relationships between social services, ethnic motivation, or crime and attacks on the news media, but here we analyze more closely each attribute and reflect on whether there are specific causal mechanisms linking the attribute and the outcome of interest.

Regarding social service provision, in previous chapters of the book social services were found to be related to both terrorism in general and the more extreme terror of attacks on schools. However, it is not clear how providing such services would be related to attacks on journalists in particular. Mechanisms connecting social services and terrorism include the notion that it can help push communities toward accepting such violence and that services might be provided to one ethnic community, which strengthens the group to attack another (e.g., Berman 2009). However, these mechanisms do not speak to the notion of attacking journalists in particular. Thus, we do not have an expectation of a relationship, and we formally state the null hypothesis.

News media H5: Social service provision is *not* associated with news media targeting.

Ethnic motivations are associated with insurgent terrorism, and this was argued to be the case because groups motivated by a particular community connection could find easy targets in civilians in a competing community. Should ethnic motivation also be associated with attacks on journalists? One way that this could be the case is if there are news agencies associated with another ethnic group. An insurgent group claiming one ethnic group could attack journalists associated with the other ethnicity. This has happened historically, but it does not seem to be common. Inductively, we could not find an example of such a motive behind an attack on journalists in our data. Overall, it is unclear how an insurgent group's ethnic motivation might lead to attacks on the news media. This again suggests the null hypothesis.

News media H6: Ethnic motivation is *not* associated with news media targeting.

Involvement in crime such as drug trafficking or extortion is associated with terrorism, and it seems possible that it is also associated with attacks on journalists. We identify two mechanisms through which crime could be linked to assaults on the news media. First, groups engaging in kidnapping or extortion could also use this tactic against journalists. The nature of journalists' work—traveling throughout the country in less secure areas, and sometimes interviewing insurgents—makes journalists prime targets for kidnapping in particular. Visiting journalists might have more money, or might be perceived to have more money, than the typical person in the area. In addition, kidnapping a foreign journalist, or one from the capital city or a pro-government news outlet, could be especially attractive to some insurgent groups. This is consistent with anecdotal evidence discussed previously, such as the kidnapping of journalists by FARC and Abu Sayyaf.

A second way that involvement in crime might be related to attacks on the news media is that involvement in crime indicates a predatory relationship with the public. These groups probably have less restraint; they are less constrained by norms against harming people in protected categories. In addition, insurgent groups that depend on economic incentives for members, such as those that come from crime, often have less control over these members (Weinstein 2006). Individual group members in this situation, even if it is not official group policy, might be more likely to kidnap or otherwise assault journalists. Through either of these mechanisms, then, crime could suggest involvement in attacks on journalists.

> **News media H7**: Involvement in crime is associated with a *higher* likelihood of news media targeting.

7.4. Empirics

This section presents a replication of the empirical models in Chapter 4, with the alternate and more specific dependent variable of terrorist attacks against news media targets. Given that the data and models were described in Chapter 3 and 4, this information will not be repeated here. The novel empirical aspect of this chapter, the dependent variable, comes from information in the GTD. To code *Attacks against journalists*, we look at all attacks associated with the groups in our data and code the variable "1" if the group carried out an attack on a news media target that year. The GTD has 22 target types, one of which is "journalists & media." As with *Terrorism* in the main models, we primarily use a dichotomous measure because we are more interested in whether a group uses terrorism or not, as opposed to whether the group uses one additional attack—for example, four instead of three. However, we do conduct additional analyses using the count version of *Attacks against journalists*.

Regarding the distribution of the dependent variable, 29 groups or approximately 21% of the 140 insurgent organizations in the data set attack journalists at some point. Groups that attack journalists usually only do this one or a few years. There are 66 group-years, approximately 5% of the sample, coded for at least one attack on a journalist. Figure 7.1 shows the distribution of attacks over time, with the cell on the left showing how many groups carry out such attacks each year and the cell on the right showing how many attacks there were per year. Both cells show a marked increase between 1998 and 2012. The variation over time in insurgent attacks on schools is somewhat similar to the variation shown with attacks on schools in Chapter 6—both

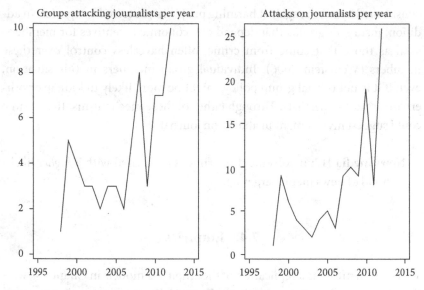

Figure 7.1 Insurgent groups' attacks on journalists over time.

increase markedly over the years. However, there are differences as well. Attacks on *schools* rocketed up in approximately 2005, peaked in 2008, and then stayed high during the remaining years. Attacks on journalists, however, increased more gradually across the years and did not peak until the final year of the sample.

7.4.1. Results

Table 7.2 replicates Table 4.8, replacing the dependent variable with *Terrorism against schools*. Model 1 contains only control variables, Model 2 includes only the key independent variables, and Model 3 is the full main model. The results have some similarities, but they are overall quite different from those shown in Table 4.8. Among the variables representing hypothesized relationships, only *Alliances* is statistically significant, which it is in both Models 2 and 3. This suggests support for this chapter's Hypothesis 3—that increased linkages with other insurgent groups is associated with a higher likelihood of attacks on journalists. Regarding substantive significance, marginal effects are plotted in Figure 7.2. The effect associated with *Alliances*, a 1 percentage-point increase in the probability of a group attacking journalists, is approximately the same as the estimated effect of

Table 7.2 Logit Models of Attacks on News Media by Insurgent Organizations

	Model 1	Model 2	Model 3
	Only Control Variables	Only Key Variables	Main Model
Carrot		0.036	0.661
		(−1.236)	(−1.25)
Stick		0.315	0.86
		(−0.404)	(−0.659)
Alliances		0.321***	0.295***
		(−0.084)	(−0.082)
Rivalry		0.225	0.024
		(−0.362)	(−0.316)
Social services		0.951	0.49
		(−0.63)	(−0.525)
Ethnic motivation		−0.669	−0.737
		(−0.641)	(−0.536)
Crime		0.215	−0.085
		(−0.202)	(−0.186)
Mixed carrot and stick	0.135		0.778
	(−0.417)		(−0.699)
Territory	1.755***		1.382**
	(−0.501)		(−0.491)
Religious	1.505**		−0.154
	(−0.598)		(−0.687)
Leftist	0.229		−0.001
	(−0.75)		(−0.66)
Group size	0.193		0.348
	(−0.342)		(−0.32)
Single leader	0.348		0.714
	(−1.318)		(−1.216)
Media attack history	0.072		0.151
	(−0.129)		(−0.155)
Conflict battle deaths	0.001**		0.000**
	(0)		(0)
State terror	0.172		0.051
	(−0.282)		(−0.276)
Democracy	0.253**		0.178*
	(−0.112)		(−0.102)

Continued

Table 7.2 *Continued*

	Model 1	Model 2	Model 3
	Only Control Variables	Only Key Variables	Main Model
GDP per capita	0.252		0.364
	(−0.269)		(−0.235)
Population (log)	−0.196		−0.1
	(−0.181)		(−0.159)
Constant	−6.133	−4.870***	−7.471**
	(−3.911)	(−0.913)	(−3.545)
N	1,240	1,240	1,240

Note: Standard errors in parentheses. Models include group random effects and year fixed effects. *$p < .10$, **$p < .05$, ***$p < .01$. This table replicates Table 4.8, in which *Terrorism* was the dependent variable.

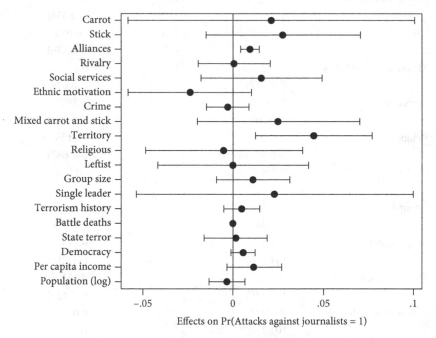

Effects on Pr(Attacks against journalists = 1)

Figure 7.2 Substantive significance of variables in the main model. Average marginal effects are shown with 95% confidence intervals.

Alliances regarding attacks on civilians generally or attacks against schools (see Chapters 4 and 6). However, the baseline probability of a group attacking journalists is lower than the baseline probability of other types of civilian targeting—only 1 or 2%. Therefore, additional embeddedness in alliances is associated with a substantial increase in the probability of an insurgent group attacking journalists.

Other variables were hypothesized to have a relationship with attacks on journalists but did not. These are *Carrot, Stick, Rivalry,* and *Crime.* It is interesting that neither of the variables measuring government actions toward insurgents have statistically significant coefficients. Although insurgents' relationships with the state apparently help explain insurgent terrorism generally, these relationships are not as helpful in explaining attacks on journalists. Rivalry is similar. It seems that interorganizational competition does not lead to attacks on journalists. Perhaps the most surprising finding is that of *Crime.* This is discussed more later.

Some variables are not associated with attacks on journalists, consistent with our expectations. The coefficients on both *Social service provision* and *Ethnic motivation* are statistically insignificant. As discussed previously, it is not clear that these group attributes should be related to journalist targeting. They are related to terrorism generally, but there is not a plausible causal mechanism that should connect them to attacks on journalists in particular. This further suggests attacks on journalists are unique in a number of ways compared to terrorism generally by insurgent organizations.

It is noteworthy that no other variables representing hypothesized relationships have statistically significant coefficients. Specifically, government concessions, government coercion, intergroup rivalry, and crime are not related to attacks on journalists. This was against our expectations. Most of the factors that explain terrorism generally do not seem to be helpful for explaining insurgent attacks against journalists, further suggesting important distinctions between terrorism in general and attacks on journalists in particular. This is discussed more later, as we consider why it makes sense that some of these factors are unrelated to assaults on the news media.

Regarding control variables, *Territory, Battle deaths,* and *Democracy* are all statistically significant and positively signed. This suggests that if an insurgent group controls territory, is in a more intensive armed conflict that year, or is in a more democratic country,[9] it is more likely to attack journalists.

---

[9] Democracy is only statistically significant at the 90% level.

This is intuitive and is discussed further later. The substantive impact of *Territory*, a 5 percentage-point increase in the probably of attacking the news media, is the largest of the model (see Figure 7.2). Regarding the other control variables whose coefficients do not achieve statistical significance at conventional levels, these results are not very surprising. For example, the lack of significance on the measure of cumulative journalist targeting incidents is consistent with the notion that most groups did not attack journalists for many years. It is helpful to take into consider this possibility because some groups, such as FARC, did attack journalists in multiple years, but this does not seem to be a trend.

7.4.2. Discussion

The results are interesting for a number of reasons. One of the factors most robustly associated with journalist attacks, and the only one correctly hypothesized to have a relationship, is *Alliances*. Interorganizational connectivity was argued to be related to attacks on the news media via two possible mechanisms: alliances proxying increased group capacity and alliances facilitating tactical innovation. Which of these mechanisms is more likely to be relevant? *Group size* is another measure often used for group capacity, and it is not related to attacks on journalists. As a result, group capacity might not explain the relationship between alliances and violence against the media. Tactical diffusion has been shown to be important for other terrorist tactics (Howitiz 2010) or terrorism generally (Polo 2020), so that could be what is occurring with attacks on journalists. Groups might learn from other insurgents the benefits of coercing journalists to report more positively about them, or of kidnapping them for ransom, and repeat this behavior. This is worthy of additional research.

Four factors were expected to be related to attacks on journalists but ultimately were not: *Carrot*, *Stick*, *Rivalry*, and *Crime*. Regarding carrots and sticks, these factors are about the relationship between the state and each insurgent group, so it seems journalists (or at least attacks on them) are somewhat orthogonal to these relationships. This might be surprising because journalists are a channel through which insurgents can communicate—for example, their "good behavior" after receiving concessions or their displeasure in response to state coercion. But apparently press releases are how insurgents communicate more often via the news media when trying to reach the government. Regarding rivalry, it does make sense that rivals use violence against each other, and perceived supporters, instead of the news media. Usually, the news media does not choose a side between insurgents, so it is not common that insurgents attack a media source for supporting their rival, as they do with pro-government media. Regarding crime, we theorized two reasons why crime might be related to the dependent variable, and we noted that some groups that engage in kidnapping (e.g., Abu Sayyaf) clearly kidnap journalists. However, this does not seem to be widespread. Another possibility is that some insurgents engaging in crime want to avoid media attention and therefore leave the media alone. Note that crime was also *not* associated with attacks on schools (see Chapter 6). Thus, crime seems to explain terrorism generally but not these specific kinds of attacks.

The control variables that are statistically related to attacks on journalists (*Territory, Battle deaths,* and *Democracy*) are intuitive. The significance of *Territory* and *Democracy* in particular could be related to the dependent variable for reasons that are relevant to the notion of embeddedness. The consistent relationship between territorial control and attacks on the press seems consistent with the idea of groups seeking to dominate the local public. Regarding territorial control, for example, other work has found it related to civilian abuse (De la Calle 2017). Groups that want control over the local population attack the news media to silence it if it is pro-government or reduce any coverage that could reduce the insurgents' influence. The examples of al-Shabaab, FARC, and other territory-holding groups discussed previously are consistent with this idea. The militants' relationship with the public, then, is essential for understanding their violence against the media.

Democracy also has some potential for contributing to the theoretical argument. Insurgents' relationships with the public are mediated through the news media in more democratic countries. As a result, the fact that journalists are more likely to be attacked by insurgents in more democratic countries makes sense. It could be that there are simply more targets in democratic countries—more journalists available. This is likely to be part of the explanation for the relationship between democratic regime type and attacks on journalists. But foreign journalists are present in civil conflict environments throughout the world regardless of regime type, including Myanmar and Syria. It makes sense that insurgents would be increasingly likely to attack the news media in more democratic countries because in these countries the public depends on the news media for information about insurgents, so insurgents coerce journalists to present them in a more positive light. The insurgent–public relationship is important, and the news media is more relevant for this relationship in more democratic countries.

7.5. Conclusion

This chapter continued empirical testing by considering how insurgent embeddedness might be related to a different kind of violence: attacks on the news media. Previous chapters focused on terrorism generally or a more extreme kind of terrorism—attacks on schools. This chapter examined violence against journalists as a specific kind of terror, one often intimately related to communication and the public. We hypothesized that some factors from the embeddedness model should be related to attacks on journalists, but perhaps surprisingly, only one was related: interorganizational alliances. Based on the results of this chapter and those of Chapters 4 and 6, alliances are the factor most robustly associated with attacks on civilians. It is interesting that each outcome—terrorism generally, attacks on schools, and attacks on journalists—has a unique causal process behind it, and thus results are quite different across models. The consistency of alliances, then, is noteworthy. Together, these findings add to the growing literature on the importance of alliances in insurgency and terrorism (Akcinaroglu 2012; Asal and Rethemeyer 2008; Bacon 2018; Christia 2012). To build upon the findings of this chapter, it would be interesting for additional research to explore case studies or social network analyses of militant group relationships and attacks on journalists to specifically explore causal pathways.

Regarding other embeddedness factors, it makes sense that indicators of relationships with the state (carrots and sticks) are not related to journalists. There is not a clear theoretical connection between these state behaviors and violence against the news media. This additional difference in findings, depending on which dependent variable is analyzed, further illustrates the uniqueness of attacks on journalists. Two variables included as controls were related to attacks on journalists: territorial control and democracy. Groups that control territory attack the news media because they depend on public support to maintain territorial control. Additionally, journalists are more relevant in democratic countries because the public plays a role in governance: journalists facilitate public exercise of political power through distribution of information. Although these were just control variables, they provide additional evidence for the role of insurgent embeddedness in explaining violence against civilians. Insurgents' relationships with the public apparently can tell us about attacks on the news media. This is generally consistent with the book's main argument that ties between insurgents and other actors explain their violence.

In general, the vast differences in results depending on whether the dependent variable is violence against civilians generally, schools, or journalists highlight the heterogeneity of "terrorism" and the divergent explanations of distinct types of attacks on civilians. Disaggregating types of attacks by insurgent groups is an essential part of understanding political violence and the ways it affects publics. Scholars have examined causal factors behind types of attacks such as suicide terrorism (Alakoc 2017; Bloom 2005; Horowitz 2010), sexual violence during civil conflict (Butler and Jones 2016; Cohen 216; Wood 2006), attacks on peacekeepers (Fjelde, Hultman, and Lindberg Bromley 2016), attacks on human rights nongovernmental organizations (Murdie and Stapley 2014), and attacks on police (Gibbs 2018), among other types of political violence. The literature is enriched by continued focus on distinct and important classes of bloodshed to better understand the broad range of political violence.

This chapter provided one of the first analyses of the determinants of insurgent group attacks on journalists. Violence against journalists is important for normative reasons, given the crucial role of journalism as a check against power, and in terms of information provision generally. Although this chapter examined violence by insurgent groups, further research could seek to understand terrorism (by any type of violent non-state actor) against journalists, whether in single case studies or globally using quantitative data. There has also been insufficient research on state and pro-government militia violence toward journalists, with few exceptions (Bartman 2018; Gohdes and Carey 2017). More research is needed on violence against journalists, but this chapter sought to offer at least a preliminary look at this important topic and show its connections to relational aspects of insurgent violence.

References

Abbas, Mazhar. "The Price of a Misleading Headline." *The Express Tribune* (Pakistan). February 7, 2013. https://tribune.com.pk/story/504159/the-price-of-a-misleading-headline.

Ahmed, Mujeeb, and Sarah Burke. "Alleged Baloch Liberation Army Militants Accused of Killing 27." NBC News. September 1, 2015. https://www.nbcnews.com/news/world/alleged-baloch-liberation-army-militants-accused-killing-27-n419326.

Akcinaroglu, Seden. "Rebel Interdependencies and Civil War Outcomes." *Journal of Conflict Resolution* 56, no. 5 (2012): 879–903.

Alakoc, Burcu Pinar. "When Suicide Kills: An Empirical Analysis of the Lethality of Suicide Terrorism." *International Journal of Conflict and Violence (IJCV)* 11 (2017): a493–a493.

Al Jazeera. "Somalia Journalist and His Wife Shot Dead." May 1, 2015. https://www.aljazeera.com/news/2015/05/somalia-journalist-wife-shot-dead-150501201809267.html.

Bacon, Tricia. *Why Terrorist Groups Form International Alliances.* Philadelphia, PA: University of Pennsylvania Press, 2018.

Bartman, Jos Midas. "Murder in Mexico: Are Journalists Victims of General Violence or Targeted Political Violence?" *Democratization* 25, no. 7 (2018): 1093–1113.

BBC. "Somalia Gunmen Shoot Dead a Journalist." May 5, 2010. http://news.bbc.co.uk/2/hi/africa/8661809.stm.

BBC. "Somalia's al-Shabab Journalist Hassan Hanafi Sentenced to Death." March 3, 2016. https://www.bbc.co.uk/news/world-africa-35716567.

BBC. "Colombia Conflict: ELN Rebels Free Dutch Journalists." June 24, 2017. https://www.bbc.co.uk/news/world-latin-america-40391221

Benítez, José Luis. "Violence Against Journalists in the Northern Triangle of Central America." *Media Asia* 44, no. 1 (2017): 61–65.

Berman, Eli. Religious Radical. *Violent: The New Economics of Terrorism.* MIT Press, 2009.

Bloom, Mia M. "Palestinian Suicide Bombing: Public Support, Market Share, and Outbidding." *Political Science Quarterly* 119, no. 1 (2004): 61–88.

Bloom, Mia. *Dying to Kill: The Allure of Suicide Terror.* Columbia University Press, 2005.

Brambila, J. A. "Forced Silence: Determinants of Journalist Killings in Mexico's States, 2010–2015." *Journal of Information Policy* 7 (2017): 297–326.

Butler, Christopher K., and Jessica L. Jones. "Sexual Violence by Government Security Forces: Are Levels of Sexual Violence in Peacetime Predictive of Those in Civil Conflict?" *International Area Studies Review* 19, no. 3 (2016): 210–230.

Carey, Sabine C., and Anita R. Gohdes. "Understanding Journalist Killings." *The Journal of Politics* 83, no. 4 (2021): 1216–1228.

Charles, Mathew. "Why Are Journalists Threatened and Killed? A Portrait of Neo-Paramilitary Anti-Press Violence in Colombia's Bajo Cauca." *Journalism* (2020, June). https://doi.org/10.1177%2F1464884920928172.

Christia, Fotini. *Alliance Formation in Civil Wars.* Cambridge, UK: Cambridge University Press, 2012.

Committee to Protect Journalists. "Attacks on the Press in 2001—Indonesia." February 2002. https://www.refworld.org/docid/47c5662911.html.

Committee to Protect Journalists. "In Mali, Rebels Assault Journalist, Force Station Off the Air." September 19, 2012. https://cpj.org/2012/09/in-mali-rebels-assault-journalist-force-station-of.

Committee to Protect Journalists. "CJP Condemns Threats by Taliban Against Afghan Media." October 13, 2015. https://cpj.org/2015/10/cpj-condemns-threats-by-taliban-against-afghan-med.

Committee to Protect Journalists. 2020. "1376 Journalists Killed Between 1992 and 2020/ Motive Confirmed." Retrieved July 17, 2020, from https://cpj.org/data/killed/?status=Killed&motiveConfirmed%5B%5D=Confirmed&type%5B%5D=Journalist&start_year=1992&end_year=2020&group_by=year.

Crenshaw, Martha. "The Causes of Terrorism." *Comparative Politics* 13, no. 4 (1981): 379–99.

Cunningham, Kathleen Gallagher, Marianne Dahl, and Anne Frugé. "Strategies of Resistance: Diversification and Diffusion." *American Journal of Political Science* 61, no. 3 (2017): 591–605.

De la Calle, Luis. "Compliance vs. Constraints: A Theory of Rebel Targeting in Civil War." *Journal of Peace Research* 54, no. 3 (2017): 427–41.

El Tiempo. "Hay 3 hipótesis del atentado: La cita ayer del personal de EL TIEMPO en Cali fue diferente a las demás." November 16, 1999. https://www.eltiempo.com/arch ivo/documento/MAM-954126.

Fjelde, Hanne, Lisa Hultman, and Sara Lindberg Bromley. "Offsetting Losses: Bargaining Power and Rebel Attacks on Peacekeepers." *International Studies Quarterly* 60, no. 4 (2016): 611–23.

FLIP (Fundación Para la Libertad de Prensa, Foundation for Press Freedom). "Dinamitada Antenna de Emisora en Puerto Asís (Putumayo)." March 5, 2005. https:// flip.org.co/index.php/es/informacion/pronunciamientos/item/1270-dinamitada-ant ena-de-emisora-en-puerto-asis-putumayo.

Gibbs, Jennifer C. "Terrorist Attacks Targeting the Police: The Connection to Foreign Military Presence." *Police Practice and Research* 19, no. 3 (2018): 222–40.

Gohdes, Anita R., and Sabine C. Carey. "Canaries in a Coal Mine? What the Killings of Journalists Tell Us About Future Repression." *Journal of Peace Research* 54, no. 2 (2017): 157–74.

Harris, Shane, Greg Miller, and Josh Dawsey. "CIA Concludes Saudi Crown Prince Ordered Khashoggi's Assassination." *The Washington Post.* November 16, 2018. https:// www.washingtonpost.com/world/national-security/cia-concludes-saudi-crown-pri nce-ordered-jamal-khashoggis-assassination/2018/11/16/98c89fe6-e9b2-11e8-a939- 9469f1166f9d_story.html.

Holland, Bradley E., and Viridiana Rios. "Informally Governing Information: How Criminal Rivalry Leads to Violence Against the Press in Mexico." *Journal of Conflict Resolution* 61, no. 5 (2017): 1095–1119.

Horowitz, Michael C. "Nonstate Actors and the Diffusion of Innovations: The Case of Suicide Terrorism." *International Organization* 64, no. 1 (2010): 33–64.

Horowitz, Michael C., and Philip B. K. Potter. "Allying to Kill: Terrorist Intergroup Cooperation and the Consequences for Lethality." *Journal of Conflict Resolution* 58, no. 2 (2014): 199–225.

International Committee of the Red Cross. "Protocol Additional to the Geneva Conventions of 12 August 1949, and Relating to the Protection of Victims of International Armed Conflicts (Protocol I). Article 79." 1977. https://ihl-databases. icrc.org/applic/ihl/ihl.nsf/1a13044f3bbb5b8ec12563fb0066f226/cbd4507e8159ebe1c 12563cd00436ec4.

Jolly, David, and Jawad Sukhanyar. "Taliban Suicide Bomber Strikes Packed Bus in Kabul." *The New York Times.* January 20, 2016. https://www.nytimes.com/2016/01/21/world/ asia/afghanistan-kabul-suicide-bombing.html.

Kalyvas, Stathis N. "How Civil Wars Help Explain Organized Crime—And How They Do Not." *Journal of Conflict Resolution* 59, no. 8 (2015): 1517–40.

Kirk, Don. "He Tells of a 2d Kidnapping by Rebels: German Journalist Freed in Philippines." *The New York Times.* July 28, 2000. https://www.nytimes.com/2000/07/ 28/news/he-tells-of-a-2d-kidnapping-by-rebels-german-journalist-freed-in.html.

Linebarger, Christopher. "Dangerous Lessons: Rebel Learning and Mobilization in the International System." *Journal of Peace Research* 53, no. 5 (2016): 633–47.

Little, David. "Not a Random Attack: New Details Emerge from Investigation of Slain NPR Journalists." *All Things Considered*. National Public Radio. June 9, 2017. https://www.npr.org/2017/06/09/531562283/not-a-random-attack-new-details-emerge-from-investigation-of-slain-npr-journalis.

Löfgren Nilsson, Monica, and Henrik Örnebring. "Journalism Under Threat: Intimidation and Harassment of Swedish Journalists." *Journalism Practice* 10, no. 7 (2016): 880–90.

Lopez, François. "If Publicity Is the Oxygen of Terrorism—Why Do Terrorists Kill Journalists?" *Perspectives on Terrorism* 10, no. 1 (2016): 65–77.

Murdie, Amanda, and Craig S. Stapley. "Why Target the "Good Guys"? The Determinants of Terrorism Against NGOs." *International Interactions* 40, no. 1 (2014): 79–102.

Obermaier, Magdalena, Michaela Hofbauer, and Carsten Reinemann. "Journalists as Targets of Hate Speech: How German Journalists Perceive the Consequences for Themselves and How They Cope with It." *Studies in Communication and Media* 7, no. 4 (2018): 499–524.

Phillips, Brian J. "How Does Leadership Decapitation Affect Violence? The Case of Drug Trafficking Organizations in Mexico." *Journal of Politics* 77, no. 2 (2015): 324–36.

Polo, Sara M. T. "How Terrorism Spreads: Emulation and the Diffusion of Ethnic and Ethnoreligious Terrorism." *Journal of Conflict Resolution*. 64, no. 10 (2020): 1916–42. https://doi.org/10.1177%2F0022002720930811.

Relly, Jeannine E., and Celeste González de Bustamante. "Silencing Mexico: A Study of Influences on Journalists in the Northern States." *International Journal of Press/Politics* 19, no. 1 (2014): 108–31.

Reporters Without Borders. "Journalist Murdered in Balochistan." February 11, 2008. https://rsf.org/en/news/journalist-murdered-balochistan.

Reporters Without Borders. "Malick Ali." n.d. https://rsf.org/en/hero/malick-ali-maiga.

Salazar, Grisel. "Strategic Allies and the Survival of Critical Media Under Repressive Conditions: An Empirical Analysis of Local Mexican Press." *International Journal of Press/Politics* 24, no. 3 (2019): 341–62.

The Star. "Hamas Crushes Fatah." *The Star* 2007. https://www.thestar.com/news/2007/06/14/hamas_crushes_fatah.html.

Weinstein, Jeremy M. *Inside Rebellion: The Politics of Insurgent Violence*. Cambridge University Press, 2006.

Wood, Elisabeth Jean. "Variation in Sexual Violence During War." *Politics & Society* 34, no. 3 (2006): 307–42.

SECTION III
FURTHER ANALYSIS OF INTERGROUP RELATIONSHIPS

8

Longitudinal Modeling
of Insurgent Alliances

8.1. Introduction

On the evening of May 30, 1972, three tourists arriving on Air France Flight 132 from Rome calmly walked through the terminal of Lod International Airport in Israel, pulled out automatic weapons, and began shooting into the assembled passengers. In moments, the three gunmen killed 26 people, including "17 Christian pilgrims from Puerto Rico, one Canadian citizen, and eight Israelis [;] 80 people were injured" (Zieve, 2012). One gunman committed suicide; another was killed (whether by his own weapon, by confederate, or by security forces has never been determined), and a third was captured (Zieve, 2012). The attackers originally claimed that they were members of "Army of the Japanese Red Star," but all were later linked to the Japanese Red Army (JRA) (Quillen 2002a, 2002b; United Press International 1972). Eventually, the Popular Front for the Liberation of Palestine (PFLP) claimed credit for the attack. As then Transport Minister Shimon Peres remarked at the time, "So far the Front has involved only Arabs and Jews. The introduction of Japanese into the picture is quite strange" (United Press International 1972).

   Indeed: Why would a Japanese communist militant group involve itself in a conflict in which it shares no territory and no ethnic or political connection? How and why would this linkage come into being? Why do groups become embedded in networks of other terrorists or insurgents? In time, the motives became clearer: Developments domestically in Japan were constraining the JRA's ability to operate (Steinhoff 1989). The PFLP opened new operational opportunities for the JRA, including the chance to engage in a spectacular attack sure to gain attention throughout the world. Maybe more importantly, the JRA needed training. With no violent peers in Japan, the JRA needed to delve into the network of communist fellow travelers to learn how to be more effective terrorists (Steinhoff 1976). The connection was made on the

Insurgent Terrorism. Victor Asal, Brian J. Phillips, and R. Karl Rethemeyer, Oxford University Press. © University of Maryland National Consortium for the Study of Terrorism and Responses to Terrorism (START) 2022. DOI: 10.1093/oso/9780197607015.003.0008

basis of shared ideological commitments, not common ethnic enmities. The quid pro quo for becoming better terrorists through co-training at a camp in Lebanon was the attack on Lod (Miller 1985; Steinhoff 1976, 1989)—an attack that was certain to have the advantage of surprise, as the quote from Minister Peres confirms.

Why, then, do insurgent organizations ally in the first place? And are these reasons stable across time, or do they change? As noted previously, ideological affinity—a common commitment to communism—provided channels through which the PFLP and JRA made common cause, but was that the only reason? Were ideological connections alone sufficient to bridge 6,000 miles and a cultural and linguistic gulf? Or is regional proximity more important? Or was the demonstrated operational capacity of PFLP enough to attract a connection from the JRA? Would that happen if one partner became less adept at killing or destruction? Or if one became particularly lethal and thus more likely to attract attention from counterterrorism forces both in its home-base country and from major powers? Although we are fairly confident that the number and type of alliance connections can significantly affect the behavior of violent non-state actors (Abuza 2002, 2003; Adamson 2005; Arquilla 1999, 2003; Asal and Rethemeyer 2008a, 2008b; Asal et al. 2009; Basile 2004; Farley 2003; Laqueur 1999; Matthew and Shambaugh 2005; Mayntz 2004; McAllister 2004; Reid, Chen, and Xu 2007; Sageman 2004; Schneider and Schneider 2002; Turk 2002), there are few—if any—rigorous, quantitative studies of both *relational formation* and *change over time.*

This chapter seeks to close this gap: What factors help explain relational formation, persistence, and change among insurgent groups? Do those factors change over time? If so, are there factors that are consistent? What seems to account for changing drivers for relational structure? To answer these questions, we must turn to a new set of tools. Although standard statistical methods are useful for studying many of the questions of interest in this book, understanding network formation, persistence, and change presents a particular set of technical problems. Fundamentally, network data do not conform to the basic assumptions of most statistical models: Network data are not derived from a random sampling process but, rather, from strenuous efforts to collect information on every member of the network. Linear statistics assume that observations are *independent* from one another. Networks are all about *interdependence* among actors. So we must seek other tools with more appropriate assumptions—stochastic actor-oriented models (SAOMs) of network formation and change.

Second, we must also decide how to approach the BAAD2 data: Should we analyze the full 15 years in a single model, or should we divide the data into periods and see how the periods relate to one another? In this instance, we will let the data guide us by attempting both. As will be shown, the data's verdict is that our 15-year study period is not one homogeneous "lump" of years but, rather, specific "epochs" that are governed by similar yet divergent rules for forming relationships between insurgents.

Section 8.2 of this chapter provides a brief, nontechnical introduction to SAOMs. Next, we delve into whether the data are best understood using a single, 15-year SAOM or multiple SOAMs, each corresponding to a particular period. Once we have established the terms of engagement with the model and data, we then turn to our findings. As will be shown, although there are some common factors that explain insurgent network formation and change, the drivers morph over time as the nature of the operating environment for insurgent organization and insurgent networks change.

8.2. Stochastic Actor-Oriented Models

To better determine why insurgent organizations made connections with one another, we used the BAAD2 data set's network and organizational data to construct SAOMs of the process.

SAOMs are a special class of statistical models that are used to overcome some of the unique challenges of analyzing social network data. The most fundamental of these challenges is that standard techniques such as linear regression are not appropriate because network data are fundamentally not a random sample. Instead, we are seeking to study a full population of a particular type of "social entity"—here, insurgent organizations. The organizations we included were not randomly selected, and without random selection the elegant mathematics of linear and other forms of regression do not apply.

Instead, SAOMs use a new set of techniques that try to model the ways organizations in a network data set are interdependent. That requires new assumptions and new techniques. The statistical models we use here start with the assumption that all members of a network may be related to one another, that those patterns of relating may have something to do with the *nature* of the organizations that are members of the network, but those patterns are also shaped by preferred forms of interaction between insurgent organizations.

To dwell on the final point for a moment, we often say that a friend of a friend is a friend or that one good turn deserves another. These folksy sayings point to expectations that we have about the way that we relate to one another as individuals or organizations. If we expect a friend of a friend to be a friend, it may say that we are more comfortable when there are no strangers in the group—that the group should be "socially closed" to strangers. If we expect a favor to be repaid with a favor, then we expect reciprocity. Broadly speaking, people and organizations have certain expectations about patterns of social interaction.

Insurgent organizations are no different. To properly model why insurgent organizations relate to one another, we have to account for these preferred forms of social interaction between insurgent organizations, sometimes referred to as "microstructures." These microstructures—reciprocity, social closure, etc.—may affect relational choices *independent* of the characteristics of two insurgent organizations that are considering working together. For example, if I go to a party and a friend introduces me to someone with whom I share nothing—we are not friends, we are from different professions, we like different football teams, etc.—I may still feel compelled to reciprocate the introduction and even spend time with that person. The social conventions of the form of interaction dictate my interactions, even if I really do not wish to engage with this stranger. A SAOM must account for these social conventions in the mathematics used to better understand why a network is structured in a certain way at a certain point in time.

To do this, SAOMs try to infer from the data what these patterns of preferred social interaction are and then use them as controls in the broader analysis of why insurgents choose to ally with one another. Microstructures, in turn, help us better understand what characteristics of organizations makes some relationships more likely and what country contexts are conducive to formation of cooperative relationships between insurgents. Collectively, then, preferred social interaction patterns, organizational characteristics, and the country characteristics of organizations that might collaborate determine how attractive any given organization is as a partner and how likely any pair of organizations is to make a cooperative relationship. We used the RSIENA package, developed by Tom A. B. Snijders and collaborators, for our analysis (Ripley et al. 2020; Snijders 2001).

Table 8.1 Number of Insurgent
Organizations Active Each
Year: 1998–2012

Year	Organizations
1998	76
1999	77
2000	80
2001	82
2002	83
2003	91
2004	94
2005	92
2006	97
2007	99
2008	98
2009	102
2010	101
2011	108
2012	108

8.3. The Data

With respect to the network data used in our analysis, for each year we collected information on the presence or absence of ties between all organizations that were active in that year. The total number of active organizations
fluctuated (Table 8.1). Although the number of insurgent organizations has
both increased and decreased from year to year, the total number in the
Uppsala Conflict Data Program (our reference for the number of active insurgent organizations each year) has tended to increase over time, with the
largest increases in 2003 and 2011.

Having established the universe of active insurgent organizations, we then
assumed that all relationships between them were "undirected," to use the
social network term. That is, we did not attempt to determine whether, for instance, Hamas reached out to Hizbullah to make a relationship or vice versa.
Instead, we simply determined whether the two organizations had a known
alliance connection or suspected alliance connection. For the purposes of
what follows, we treat both alliances and suspected alliances as evidence of

a "tie"—a collaborative connection between insurgent organizations that we wished to model. In social network terms, then, our "dependent network" (the data we wished to model and explain) was undirected and dichotomous.

We then sought to understand what factors helped explain the presence or absence of those ties by modeling the microstructures (reciprocity, social closure, etc.), organizational characteristics (ideology, membership size, political participation, etc.), and country variables (wealth, regime type, use of political terror against citizens, etc.) that help predict the presence or absence of the ties we found between organizations in our yearly data and how those patterns changed over time.

8.4. Modeling the Data: 15 Years or Three Periods?

The original plan was to build a single model using data from all 15 years in order to better understand the enduring factors that explain insurgent networks and their patterns of change. In fact, we did just that: We were able to model all 140 organizations and 15 years in a "grand unified model." SAOMs have several standards for whether they are adequate representations of the data and the processes that lead to the results. Our 15-year model met the first and possibly most important indicator of model adequacy: convergence. Convergence is said to occur when a SAOM's estimated coefficients can reproduce the underlying data with high fidelity across thousands of simulated networks. Our 15-year model could do that with a very high degree of fidelity indeed: Our final model met the accepted criteria that all t-value measure of difference between simulated data and actual data was less than 0.10. That is, we were sure to a very high degree of certainty that the average result from our simulated networks could reproduce the major features of our actual data very precisely.

However, there are other measures of model adequacy for SAOMs. One of those is time heterogeneity. Here, our 15-year model did not meet the mark. Time heterogeneity is when there is evidence that the relationship between a factor—for example, organizational age or ideological commitments—and network tie formation or persistence does not stay the same across the entire period. For instance, it could be that long-established insurgent organizations prefer to connect to other well-established insurgents through the 1990s, whereas a decade later older organizations are equally likely to connect to old or young organizations. When we modeled all 15 years together, we implicitly assumed that our factors affected insurgent network structure and tie formation the same way and at the same strength in 1998 as in 2012. However, our tests demonstrate that this assumption is wrong.

Instead, our tests suggested that the following could be true about our network data and how insurgent organizations create ties with one another:

- Some variables might affect tie formation in one period but not in others (i.e., the estimated coefficient may be zero in some years or periods but statistically significant in others); and/or
- The *sign* of the coefficients on a variable could differ from period to period even if the variable is significantly related to network formation; and/or
- The *size* of the coefficient and thus its effect could differ from period to period, even if the coefficient is statistically significant *and* has the same sign period to period.

After consulting with Tom A. B. Snijders (the progenitor of SAOMs and the RSIENA estimation package), we decided the best strategy was to divide the data into shorter periods and model each separately. This would allow some variables to fall out of the model, change sign, or be of greater or lesser valence (i.e., coefficient size) between periods. Ideally, the periods should be at least 3 or 4 years and should correspond to known events or environmental changes that could plausibly cause variables to matter more, less, or differently from one period to another.

The most natural dividers in the 15 years of data we are analyzing relate to changes in the U.S. engagement with terrorism and insurgency. As we reviewed the data and period, the following subdivisions made the most sense substantively:

Pre-September 11, 2001 (9/11) (1998–2001): In this period, the United States viewed terrorism as a significant and potentially growing threat, especially after the 1998 bombing of the U.S. embassies in Kenya and Tanzania and the 2000 attack on the USS Cole (see National Commission on Terrorist Attacks upon the United States, 2004, Chapter 6) but did not have substantial overseas deployments tasked primarily with countering the threat to U.S. interests at home and abroad from the actions of terrorists and insurgents. U.S. attention to the activities of insurgent organizations was sporadic, reactive, and primarily focused on threats in the Middle East.

Post-9/11 (2002–2007): The second period we define as from the September 11th terrorist attacks to the Bush administration's surge in Iraq in 2007. This period was characterized by the massive U.S. response [coordinated with the North Atlantic Treaty Organization (NATO) and other allies] to the 9/11 attacks, leading to the invasion of Afghanistan and later Iraq. Although the focus of counterinsurgency (CI) and anti-terrorism (CT) activity was in the Middle East, U.S. policy embraced the concept of a "global war on terrorism" (Jenkins, 2017) that included widespread efforts to disrupt terrorist and insurgent organizations inside as well as outside the Middle East. U.S. and allied responses to terrorism and insurgency relied very heavily on coercive actions centered in military operations. The zenith of the "hard" coercive response was the Bush administration's surge in Iraq in 2007.

Post-surge (2008–2012): This period, which ends with the U.S. formally removing most of its remaining soldiers from Iraq and the rising influence of the Islamic State, is marked by a change in the U.S. approach to counterterrorism and counterinsurgency. The United States adopted what some have characterized (Gilmore 2011) as a "classic" counterinsurgency strategy, which included efforts to engage with local populations and reduce reliance on purely coercive strategies. Although now disputed with respect to its importance, this was also the period in which the so-called Sunni Awakening in Iraq played a role in U.S. efforts to disengage in Iraq and was a marker of the new approach that embraced a classic "hearts and minds" strategies. Finally, the election of Barack Obama on a platform that deemphasized U.S. troops on the ground—especially in Iraq and to a lesser extent Afghanistan (Obama 2008)—as a means of prosecuting counterinsurgent strategies firmly entrenched the idea that the United States was reluctant to engage more heavily with troops on the ground. Instead, the United States would seek to pursue strategies that engaged with adversaries and used "remote control warfare" when possible through new capabilities such as unmanned aerial vehicles or "drones" (Byman 2013).

Fundamentally, there were strong reasons to believe that insurgent organizations would act somewhat differently during these three periods, leading to differences in how insurgent networks would form and change over time. And those changes, in turn, would show up in our modeling of the BAAD II data as differences in factors that influence the likelihood that organizations make, break, or sustain ties with one another.

8.5. Findings: Understanding the Determinants of Insurgency Network Structure

8.5.1. Factors That Are Consistent Across the Study Period

Like standard linear regression, the standard used in SAOMs for determining whether a factor is statistically significant is the t-statistic associated with the factor's coefficient. As is usually the case, the t-statistic measures whether the coefficient is so large that it cannot be the result of random chance. By convention, factors with a t-statistic of 2.00 or greater are said to be statistical significance in SAOMs. Here, 2.00 corresponds to 95% confidence. Effects with t-statistics greater than 1.666 are significant with a 90% certainty and are sometimes considered weakly significant. We note levels of significance across our findings.

Our analysis found four factors with t-values above 1.666 across all three periods that we modeled.

8.5.1.1. Preference for Relatively Sparse Networks (Density)
In all four periods, the *density* coefficient is negative and statistically significant ($t > 2.00$ for all year). In general, insurgent organizations are selective in their creation of connections: There are substantial risks to making connections (a friend of a friend may be the Central Intelligence Agency) and costs. Costs can include not only the sharing of scarce physical resources but also the reputational harm that may come from a known ally committing a taboo act (use of suicide terrorism) or looking weak if an attack fails spectacularly.

Far less than 50% of all possible ties in the network are actually realized, and in each period there are a substantial number of isolates. In fact, 19 of the 140 organizations (13.5%) are operative for the entire 15 years for which we have data yet never make any connections during any of those years. During the pre-9/11 period, approximately half of all operative organizations (42) were isolates over the 4 years; during the surge years, approximately one-third (34) of all active organizations were isolates; and during the post-surge period, 40 active organizations made no ties.

Thus, the norm was for most organizations to have relatively few ties, but with definite exceptions. Al-Qaeda, for instance, averaged 10 connections in each of the 15 years; the Taliban averaged more than 5 connections in every year; Al-Aqsa Martyrs' Brigade averaged more than 4 connections in every year; and organizations founded more recently, such as Tehrik-i-Taliban Pakistan and Al Shabab, averaged more than 3 connections per year during their relatively shorter life spans (6 years and 7 years, respectively). Some organizations clearly choose to make broad alliances as a strategic choice, but for many, their relationships with other insurgents are few, if any.

8.5.1.2. Preference for a Friend of a Friend to Be a Friend (Transitive Triads)

Transitive triads are a marker of social closure—a preference for everyone in a group to know one another. The evidence on the preference for social closure is weaker in these data than in our work on terrorist organizations (Asal et al. 2016), with the coefficient on transitive triads only weakly significant pre-9/11 ($t = 1.771$) and post-9/11 ($t = 1.787$) before becoming strongly significant post-surge ($t = 5.125$). Yet there is strong reason to believe insurgent organizations prefer to make connections where all members of the group know one another. Closed networks make it easier for people and organizations to monitor one another: If I have two friends and we all know each other, we can mutually monitor each other and report "defections" to one another. Network research has consistently demonstrated that closed networks tend to be high in trust generally and can help reinforce trust through monitoring (Burt 2005; Gill et al. 2014; Lee, Park, and Rethemeyer 2018). When operating in an illicit environment in which trust is strained at best, knowing that everyone else is known to each other provides a basis for trust and cooperation.

8.5.1.3. Preference to Make Relationships with Organizations in the Same Country (Same Country)

In all three time periods, organizations preferred to make relationships with other organizations within the same country, although the evidence post-surge is slightly more equivocal ($t = 1.822$). There are costs to making relationships; proximity reduces cost. The closer one is, the less money is spent on trains, planes, and automobiles and the less time is lost in transit. Organizations in the same country have more secure options for coordination of effort, including direct, face-to-face communications, simple courier systems, and use of electronic communications that do not use the internet as the delivery mode (e.g., USB drives passed hand to hand). Crossing international borders creates a panoply of risks, so allying within one country is preferable. Finally, organizations may share goals within the same country, which makes internal alliances attractive. However, rivalry for the same set of adherents (e.g., the Palestine Liberation Organization and Hizbullah) may make same-country alliances less attractive in some instances. On balance, however, insurgent organizations tend to strongly prefer to make alliances with other organizations within the same home country.

8.5.1.4. Organizations That Provide Social Services to Individuals in Territories They Control Make/Attract More Connections (Social Services)

Why is this so? There are a couple of possible explanations. First, social service provision may be a proxy for capacity. Providing social services is expensive and difficult. Provider organizations must be better organized, structured, and funded. This makes the organization more attractive as a partner.

Better organized and resourced organizations may also be more willing to take a risk on new relationships. Starting from the premise that social service providers are better resourced and managed, such organizations can weather a bad choice once in a while. In other words, social service providers may be large enough to be less risk averse and thus open to taking chances on relationships that are high risk/high reward. Although we found no evidence that organizational size (another proxy for capacity) was a significant factor in network structure or formation during the first two periods (but did find some evidence for a size effect post-surge), it could be that social service provision is a more consistent marker of organizational competence and capacity than simple counts of members.

Second, social service provision may also be a marker of success. Social service provision is generally a governmental function. Providing these services is a mark of sufficient control that the organization is able to function at some level like a state rather than an organization. Insurgent organizations are seeking to supplant the incumbent state for some span of territory. Providing social services is a marker that the insurgency has actually succeeded. Thus, other organizations may seek relationships with social service–providing insurgents in order to learn how they have managed to convert themselves into a state or at least a state-in-waiting.

8.5.2. Factors That Are Consistent for Two Periods: Pre-9/11 (1998–2001) and Post-9/11 (2002–2007)

There is one factor that remained consistent from 1998 to 2007: the tendency for terror users to connect at a higher rate. The measurements of "terrorist activity" that best fit the data were slightly different by period. For the pre-9/11 period, the number of terrorist incidents fit the network data best. For the post-9/11 period, the number of fatalities fit best. Nevertheless, the story was the same: The more terrorist activity an insurgent organization engaged in, the more connections it tended to consummate.

Why would organizations that engage in terrorism tend to create and receive more connections? There are at least three possible explanations.

8.5.2.1. Diffusion

One explanation is that terrorism during the pre-9/11 and post-9/11 periods was an emerging technique that groups wished to learn from one another—an argument based in the concept of diffusion (e.g., Polo 2020). It could be that insurgents that *were not* using terror were reaching out to insurgents that *were* using terror to learn.

On the face of it, this is an attractive argument, especially if we focus on the aggregate data on terrorism use pre-9/11. Table 8.2 shows that pre-9/11, the number of organizations using terrorism increased, the average number of incidents per organization increased, and the proportion of organizations that were active also increased. Something was driving this increase; it could be that organizations were learning from one another.

However, the pattern begins to fall apart during the post-9/11 period. Only approximately one-third of organizations used terror between 2002 and 2007, so terror use declined during the post-9/11 period. But then the pattern reversed again: After 2007 (post-surge), the use of terror skyrocketed. On average, 45% of active organizations used terrorism each year during the post-surge period (Table 8.3). The post-surge period is the highest period of terror use, and the average number of incidents is substantially higher. Yet, our analysis did not find that terror-using insurgent organizations were likely to make or receive more connections in the post-surge period.

Moreover, when we looked more closely at the data, the evidence for diffusion as an explanation became even thinner. To understand why, we have to first define what would constitute diffusion. There are several logical prerequisites and then some assumptions. First, as a matter of logic, diffusion

Table 8.2 Aggregate Terror Incident Statistics by Year

Year	Average No. of Incidents per Group	Median No. of Incidents	Groups with > 1 Incident	Percentage of Groups Using Terror	Groups Active This Year
Pre-9/11 Period					
1998	2.00	0	24	0.32	76
1999	3.01	0	29	0.38	77
2000	4.29	0	33	0.41	80
2001	4.41	0	38	0.46	82
Post-9/11 Period					
2002	3.93	0	30	0.36	83
2003	3.46	0	28	0.31	91
2004	2.28	0	31	0.33	94
2005	3.50	0	31	0.34	92
2006	3.61	0	28	0.29	97
2007	3.95	0	35	0.35	99
Post-Surge Period					
2008	10.17	1	53	0.54	98
2009	11.72	0	49	0.48	102
2010	12.46	0	40	0.40	101
2011	9.56	0	44	0.41	108
2012	12.81	0	48	0.44	108

Table 8.3 Period Averages for Insurgent Groups Using Terror and Incidents per Year

	Average Percentage of Active Groups Using Terror	Average No. of Incidents
Pre-9/11 (1998–2001)	39.2	3.43
Post-9/11 (2002–2007)	33.0	3.45
Post-surge (2008–2012)	45.4	11.34

cannot have occurred if the organization does not use terrorism, has used terrorism in the past, or has no connections to terror-using insurgent organizations. Second, if an organization uses terrorism in its founding year or in the year immediately after founding, we would not say that terrorism as a tactic diffused to that organization, even if it has alliances with terror-using insurgents. Use of terrorism at "organizational birth" points to the use of terror as a founding principle and tactical preference rather than a practice picked up from others. So just how much diffusion seems to have occurred, given these constraints on our definition?

The answer for both pre-9/11 and post-9/11 is *vanishingly little*. To provide some context, of the 140 organizations that are in the BAAD2 Insurgent data set, 43 had used terrorism before 1998, or approximately 31%. So 97 organizations could potentially have been "infected" with terrorism by other insurgent organizations. During the pre-9/11 period, only 14 organizations used terrorism for the first time. Of these new terror users, 7 used terror at founding or in their first year after founding, 2 had no ties at all, and 2 more had ties only to non-terror users. Thus, only 3 organizations may have "taken up" terrorism due to diffusion: the Communist Party of Nepal–Maoist from the Maoist Communist Center, Islamic Movement of Uzbekistan (IMU) from al-Qaeda, and God's Army from the Karen National Army.

The post-9/11 period also has few candidates for diffusion. More organizations adopted terror in this period: 26. But of these, 13 used terror in their founding year or the year following, and 8 more had no ties at all. Two new users had ties, but not to terror users. Which again leaves 3 organizations that might have been candidates for diffusion: Black Widows from the National Democratic Front of Bodoland, the Islamic Jihad Group from al-Qaeda, and the Ogaden National Liberation Front from the Oromo Liberation Front.

By contrast, post-surge, a similar accounting finds that of the 25 new terror users between 2008 and 2012, at least 12 could have taken up terrorism due to diffusion from other terror users. That is, 12 organizations were new terror users, these organizations had ties to terror users, and these 12 organizations took up terrorism at least 2 years after founding. So, post-surge, there are many opportunities for diffusion, but our modeling found no evidence that use of terror was an important factor in network relationship building and change. In fact, our models of pre-9/11, post-9/11, and post-surge network structure and formation are exactly the opposite of what a diffusion model would predict. So what else might account for relationship formation being built on the basis of use of terrorism?

8.5.2.2. Birds of a Feather
Another possibility is that organizations that use terror build relationships with other terror users. Terrorism is a significant norm violation. Most societies view killing of civilians as a significant violation of ethical and moral obligations. It may be that terror users and non-terror users would sort into

separate groups. Thus, organizations may prefer homophilous connections with respect to this activity: The non-terror users may not wish to stain their own legitimacy by using a norm-violating tactic, thus pushing the terror users together. However, SAOMs allow for direct assessment of this hypothesis through the inclusion of a parameter to determine if organizations prefer connections to other terror users. Our modeling found no evidence in any period that terror users preferentially connected to terror users or that non-terror users preferred non-terror users.

8.5.2.3. Terrorism as Marker of Success

A third possibility—and the most likely explanation—is that terror users are viewed as successful. Although organizations may not necessarily need to learn terror as a technique—killing civilians is easier than killing trained and armed police, paramilitary, or military personnel—they could view terror users as successful. Terror attracts local, national, and even international media attention, making terror users more salient as potential partners. Terror-using insurgent organizations are sufficiently resourced and organized to engage in at least two forms of violent political activity: killing of uniformed representatives of the state and civilians. Finally, terror-using insurgents are successfully navigating two "gaps" that may be critical to insurgency success (Eizenstat, Porter, and Weinstein 2005). Using terrorism increases the security gap by instilling fear among civilians and demonstrating the incumbent state's inability to protect. On the other hand, the insurgent organization through its actions may be widening the legitimacy gap, insofar as attacking civilians is generally taboo and thus makes the insurgent's task of being accepted as a credible replacement for the incumbent state more difficult. Regular users of terror are navigating this dilemma successfully, which may in turn make them sought-after alliance partners.

8.5.3. Factors That Are Consistent for Two Periods: Post-9/11 (2002–2007) and Post-Surge (2008–2012)

Four factors became consistently important after 9/11 through to the end of the post-surge period:

- Organizations preferentially connected not only with organizations in their same "home base" state but also in those that were contiguous with the home base state.

- Organizations inspired by an ethnonationalist ideology preferred to make alliances with organizations sharing that ideological orientation.
- Older organizations tended to make fewer connections.
- Organizations that were state sponsored—that is, those that were receiving direct support from a state in the form of financial aid, advice, weapons transfers, training, etc.—tended to make or receive more relationships.

Unlike the previous discussion of terror as a tie-formation motivator, these four post-9/11 tendencies are relatively straightforward. First, the logic of making alliances with insurgent organizations within the home base country applies straightforwardly to making alliances with organizations based next door. The motivation to look beyond the home base after 9/11 may have several roots. Organizations may have needed to find safe havens outside their main operating area as the United States directly and indirectly sought to suppress all forms of terrorism and insurgency. Although increased CT and CI efforts in some regions drove organizations to seek friends more widely, U.S. pressure on governments to step up CT and CI efforts in their sovereign territories may also have played a role.

More speculatively, the increasing availability of internet-based communications in even the most inaccessible corners of the world may have also reduced the costs of building alliance connections beyond the home base. By the post-surge period, increased cell coverage, satellite services, and internet availability made it easier to create and maintain alliance relationships with wider geographic scope. According to the International Telecommunication Union (2017), between 2005 and 2010 (which, respectively, fall in the middle of the post-9/11 and post-surge periods), internet usage increased from 16% to 30% globally, with the increase in the developing world from 8% to 21%. By region, the proportion of the population who were internet users increased from 2% to 10% in Africa, 8% to 26% in Arab states, and 9% to 23% for citizens of Asia and the Pacific (International Telecommunication Union 2017). Changes in technology and greater knowledge of how to use this technology safely (meaning without being caught by U.S. and other intelligence services) changed the definition of "local" after 9/11.

The preference for connection to co-ideologists is ultimately an expression of homophily, one of the most durable reasons for relational formation. In our earlier work (Asal et al. 2016), we found strong evidence that terrorist organizations form relationships along ideological lines. It was somewhat surprising not to find the same in the pre-9/11 period in the insurgent network, but after 9/11 shared ethnicity-based ideology is a strong basis for tie formation. In Figures 8.1–8.3, the nodes in deep blue, light green, green, and black all have an ethnonationalist component to their ideology. Organizations represented by deep blue nodes are ethnonationalist; light green represents those that are ethnonationalist and leftist; green nodes represent those that are ethnonationalist and religious; and black represents organizations that are ethnonationalist, religious, and leftist.

A quick glance at the figures shows what the modeling also revealed: There are few linkages between ethnonationalists pre-9/11 but many more thereafter. As we note in the next section, the increasing number of red nodes—those inspired by religious (and here, mostly Islamic) ideologies—becomes a prominent feature of the post-surge network.

Shared ideology is a strong basis for insurgent relational construction because it serves to align goals and provide a common framework for interpreting facts and actions, common touchstones for what is acceptable and unacceptable behavior, and (as we have argued previously; see Asal and Rethemeyer 2008a) a common way to identify "us" and "them" as a guide to which persons are entitled to protected status and which may be killed in the service of political goals.

During the pre-9/11 and post-9/11 periods, older organizations generally made fewer connections. Whether this is because older organizations were reluctant to make connections or whether older organizations were less "popular" and thus less sought out than their "younger" peers, we cannot say, because this data is not directional. In previous work (Asal et al. 2016), we found that older *terrorist* organizations are more popular and more sought out as alliance partners. Insurgency seems to be different. In part, many insurgencies are older and better established. In other areas of research (Hatmaker and Rethemeyer 2008), there is evidence that older organizations settle into long-term patterns of interaction that are not easily interrupted. Organizations, like people, create and sustain stable sets of relationships in order to help sustain operations (Hatmaker and Rethemeyer 2008; Pfeffer 1987; Pfeffer and Salancik 1978). Organizations in particular accumulate investments in regular relational interaction that are costly and difficult to replicate. As Oliver Williamson (1981) discovered, these transaction costs cause many organizations to stay in static, even harmful, long-term interactions because the costs of breaking old relationships and building new ones are substantial. For instances, organizations may buy the same weapons so that cross-training is possible or use the same apps and communications technologies to ease coordination. If the partnership dissolves, the benefits

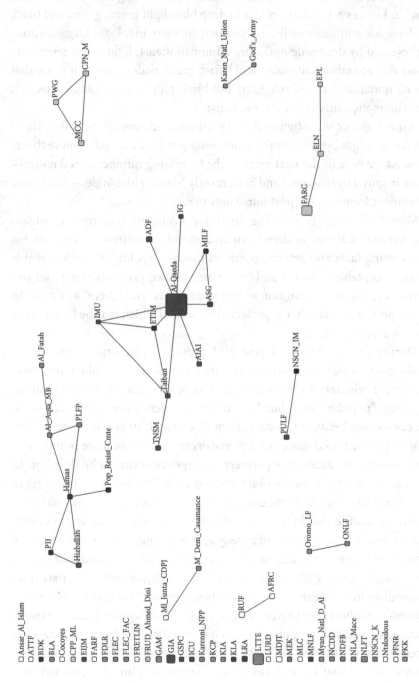

Figure 8.1 Insurgent network, 1998–2001.

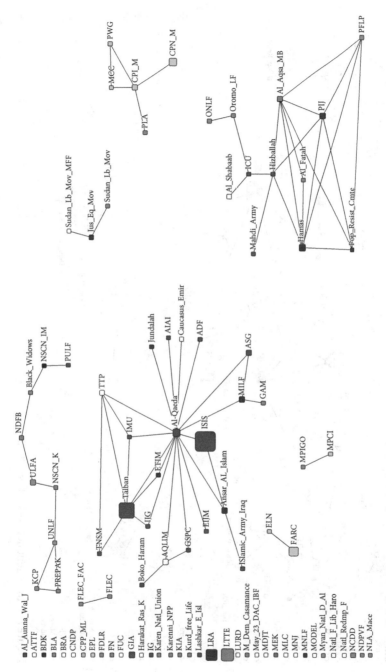

Figure 8.2 Insurgent network, 2002–2007.

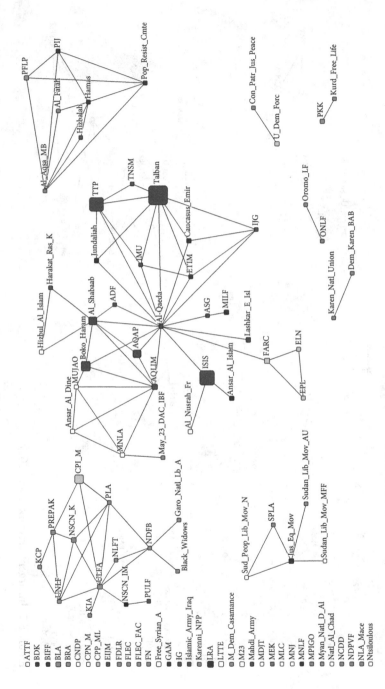

Figure 8.3 Insurgent network, 2008–2012.

of those choices and the sunk investment are lost to both parties. The jilted ally may impose a cost on breaking the friendship. Thus, organizational "divorces" have costs just like personal divorces (Hatmaker and Rethemeyer 2008). Second, stable friendships are safer. Insurgents know whether their existing peers are likely to be infiltrated by intelligence, paramilitary, or rival insurgent organizations. Making new relationships risks mistakes that lead to infiltration. Finally, established organizations can choose to turn down partnerships more selectively. Well-established organizations do not need friends and partners as badly as do younger, less established organizations.

The final commonality is that organizations with state sponsorships are more likely to form alliance relationships. Previous work has also found evidence of this (Asal et al. 2016; Popovic 2018). Post-9/11, having a state sponsor might have provided important advantages when seeking to survive the U.S.-led effort to prosecute the "war on terror." State sponsorship gives insurgent organizations a particularly deep base of resources, training, and intelligence with respect to CT and CI efforts by the home base state and its international partners, such as the United States. State-sponsored organizations are also well poised to broker relationships between the sponsor and other organizations that may be seeking sponsorship. Thus, being sponsored may make an organization attractive or "popular."

Being sponsored may also lead to matchmaking by the sponsor. In our previous work on networks and terrorist organizations, we found a particularly potent example of this in Afghanistan, Pakistan, and Kashmir (Asal and Rethemeyer 2007). In the immediate aftermath of the U.S. invasion of Iraq, the Pakistani Inter-Services Intelligence agency (ISI) rapidly developed connections to all of the insurgent and terrorist organizations in the region as a way to help protect its investment in organizations that fostered its interests in the northern and eastern frontiers of Pakistan. In the course of building those relationships, ISI also helped build relationships among the organizations with which it worked. The sponsor itself, then, helped facilitate relationships between its proxies.

8.5.4. Factors That Appear, Disappear, and Reappear: Pre-9/11 and Post-Surge

Our analysis also found two factors that were important pre-9/11, disappeared post-9/11, and then reemerged during the post-surge period: exposure to violent counterterrorism efforts and efforts to traffic in drugs (although the importance of drug trafficking pre-9/11 is less clear because the *t*-value was lower: 1.779). What might account for this on again/off again pattern of relational formation?

Turning first to violent CI/CT, during the pre-9/11 period, being subject to violent CI/CT led to more tie formation. During this period of relatively low attention to insurgency and terrorism—and especially little coordinated effort by the United States on a global scale—being subject to violent CI/CT may have acted as a marker of success: "See, we are so prominent that they pay attention to us in the form of violent efforts to suppress us." As we have suggested previously, organizations may seek out peers and mentors that are more successful, so before 9/11 being suppressed may have been a good indicator that an organization was being successful and thus was an attractive alliance partner.

After 9/11, the U.S.-led CT effort uniformly embraced violence as a strategy for CT/CI. In this environment, violent suppression may have had little utility as a marker of success. To be an insurgency organization was to be subject to violence, often from the United States and allied forces. Thus, we found no effect on relational formation from being subject to the "stick" form of CT/CI efforts.

After the surge, the pattern changed. Now our modeling found that those that were "sticked together stayed together." That is, organizations subject to violent suppression were more likely to connect with each other. Why the change? First, U.S. and allied effort at CT and CI became more nuanced. As noted previously, the U.S. switched to a more "traditional" and balanced CT/CI strategy, meaning that violence was used more selectively. Violence as a marker of behavior once again had meaning.

Second, the nature of the violence changed. As the United States drew down its ground-based operations in Iraq and Afghanistan, the Obama administration turned increasingly to drone strikes. According to the Bureau of Investigative Journalism (2020), drone strikes in Pakistan increased from 37 in 2008 to 128 in 2010 but declined to 50 in 2012. As Jenkins (2017) noted,

> While the Obama administration was wary of committing ground forces, it was not reluctant to take out terrorist leaders. Obama risked the raid that killed Osama bin Laden. He also oversaw a tenfold increase in the targeted killings of terrorist leaders and cadre that Bush had initiated. Special

operations and airstrikes became the principal expression of America's counterterrorist strategy.

Fundamentally, insurgent organizations needed better information in this environment: about U.S. and coalition intentions, best practices to avoid airborne threats, and tactics to avoid infiltration by human and electronic assets. Obama was also more selective in his use of force, symbolized by his retirement of the term "global war on terror" in favor of "countering violent extremism" (Jenkins 2017). In short, organizations that were literally still in the crosshairs in 2008 and beyond had something in common and a reason to collaborate through an alliance.

Turning now to the drug trade, the pre-9/11 and post-surge periods could not look more different. Before 9/11, drug traffickers tended to be less sought out as alliance partners, but post-surge, drug trafficking was a basis for building a larger portfolio of alliance connections. Again, why the difference? As is well known, the Taliban, Hizbollah, Kurdish Workers Party, and IMU all engaged in drug trafficking as a substantial source of funding in the 1990s. Our 15 years of data suggest that there are sets of organizations that have consistently used trafficking as a financial strategy, with 8 organizations using trafficking in all three periods and in at least 10 of 15 years in our data: Abu Sayyaf Group, al-Qaeda, Hizbollah, IMU, the National Liberation Army of Colombia, the Revolutionary Armed Forces of Colombia, Shining Path, and the Taliban. These 8 organizations were the core group of traffickers before 9/11.

Nevertheless, drug trafficking was not widely used at this point in time, with only 13 of 76 organizations actively engaging in drug running. In addition, a number of groups that emerged in the 1990s were Islamically inspired. Most Islamically inspired groups ruthlessly suppress all use of drugs among adherents, and in these early days organizations were not willing to create a wall between their commercial and ideological practices (the examples just cited notwithstanding). Most notably, the Taliban forbade all drug use and experimented with a ban on opium production for a time in 2000 and 2001 (Harding 2001). Drug trafficking was not a skill that many of the Islamically inspired organizations wished to learn, and those that engaged in it were well-established as drug merchants. Of the 13 drug trafficking organizations pre-9/11, 8 would go on to traffic in almost every year between 1998 and 2012, and only 2 (the Kosovo Liberation Army and West Side Boys) engaged in trafficking for as few as 2 years across our study period. Pre-9/11 traffickers did not need to learn from one another, and the new insurgencies built during this period were often ideologically opposed to trafficking.

During the post-9/11 period, drug trafficking did not appear to play a role in structuring networks one way or the other: Our models found no effect. In part, this reflects the fact that although 10 organizations engaged in trafficking for the first time between 2002 and 2007, 8 of those 10 organizations engaged in trafficking in only 1 year. There was insufficient engagement with trafficking by new organizations to build a basis for relational formation focused on an interest in drug trafficking.

However, by the post-surge period, our models found that drug traffickers tended to have more connections than those organizations that did not engage in drug trafficking. During the post-surge period, 10 organizations took up the drug trade, but unlike the post-9/11 period, 5 of these 10 organizations used this funding method for at least 2 years. In addition, 4 organizations that trafficked for only 1 year during the post-9/11 went on to traffic for at least 2 years in the post-surge period. A number of prominent insurgent organizations had established themselves firmly as traffickers by this period, and many of these were Islamically inspired. The Taliban, ISIS, and al-Qaeda had all established the principle that Islamically inspired organizations could engage in external trafficking even as they brutally punished adherents who used drugs (for a discussion of ISIS' approach to drug prohibition, see Tharoor 2016). Finally, there are reasons to believe that drug trafficking had become more important to the overall financial flows that supported insurgency, although the evidence is hotly debated (e.g., see Daly 2016; Omelicheva and Markowitz 2019).

Drug trafficking, then, could be a basis for relational formation. First, drug trafficking inherently requires movement of substances outside primary operational areas for insurgency organizations. Operational alliances help ease the movement of drugs. Drug networks are often interorganizational in nature: To be successful, drug traffickers need to have operational alliances both to other like-minded organizations and to criminal enterprises that dominate the drug trade (e.g., see Kenney 2007; Malm and Bichler 2011).

Second, drug trafficking has become an increasingly sophisticated operation that leverages technology effectively to coordinate and facilitate growth, processing, and sales globally. Like many economic endeavors, drug production is a knowledge- and data-intensive business, as the Executive Director of the United Nations Office on Drugs and Crime noted in 2017 (Fedotov 2017). Established traffickers had knowledge and connections that made them valuable as alliance partners. Newly formed insurgent organizations had financial needs that could be best met through drug trafficking, but they needed help to get established. Alliances built around common interest in the drug trade resulted in the post-surge period.

8.5.5. Factors Unique to Particular Periods

Each of our three periods had some relational factors that were unique. The period-specific factors are concentrated in the findings for the post-9/11 period and the post-surge period. Each is characterized by a set of relational drivers unique to the political and CT/CI context of the time.

During the post-9/11 years, two interlocking effects were important during this period of maximum U.S. CT/CI effort: the role of political terror and the special character of isolates. Turning first to political terror, we merged into the BAAD2 Insurgent data set Amnesty International's Political Terror Scale for each insurgent organization's home base country, as found in the Quality of Governance data set (Teorell et al. 2016). The Amnesty International measure has five levels, with 1 being "Countries under a secure rule of law, people are not imprisoned for their view, and torture is rare or exceptional. Political murders are extremely rare" and 5 being "Terror has expanded to the whole population. The leaders of these societies place no limits on the means or thoroughness with which they pursue personal or ideological goals" (Teorell et al. 2016, 294). Our modeling for the post-9/11 period found a clear disincentive to relational formation as more political terror was applied. In the post-surge model, organizations located in countries with higher levels of political terror were significantly less likely to form additional relationships.

The reasons for this finding are complex. One possibility is that political terror is a stand-in for violent forms of CT/CI. However, the correlation between "stick" (i.e., violent) or "stick & mixed" (violent and nonviolent) CT/CI strategies is very low: 0.046 and 0.163, respectively. So it seems unlikely that political terror is a proxy for violent suppression of terrorism and insurgency.

Instead, this may reflect a particular period of reaction to the 9/11 attacks. As other research has found, the United States and its allies—mostly developed countries, many in NATO—became more open to political terror. The average score on the Amnesty International scale increased for close allies of the United States (Goderis and Versteeg 2012). As is also well-known, the United States engaged in enhanced interrogation techniques and utilized "black sites" and third countries for these practices (Barnes 2016), culminating in the atrocities documented at the Abu Ghraib facility in Iraq that were justified, in part, by legal frameworks created by the Bush administration (Greenberg and Dratel 2005). Political terror was a form of CT/CI strategy that flowed out from the United States and was embodied in U.S. policy. Our evidence suggests that it was successful as a means of inhibiting relational formation among insurgent organizations.

The exact mechanism is difficult to discern solely with the BAAD2 data. It may be that the use of political terror (even if in somewhat attenuated form) by countries that since the Nuremberg Trials had promoted universalist approaches to human rights reflects the intensity of CT/CI effort across multiple dimensions—not just violent police/military efforts against organizations. Countries subject to high levels of political terror were likely engaged in very broad campaigns of suppression, backed by the full support of the United States in the aftermath of 9/11.

During the post-9/11 period, the character and nature of being an isolate also changed. In the pre-9/11 and post-surge periods, there was no particular tendency for isolates to stay isolates. But during the post-9/11 period, there was a pronounced tendency to stay an isolate if one was already. This probably speaks to the enormous pressure that U.S.-led CT/CI efforts put on terrorist and insurgent organizations. As noted previously, during this period the Bush administration styled its CT efforts as a "global war on terrorism" backed by military operations not just in the Middle East and North Africa but also in South Asia and even the Philippines against Abu Sayyaf (Stentiford 2018). Alliance building could be quite dangerous because the United States interpreted any connections as potentially "toxic."

During the post-surge period, the tendency for isolation to be a stable state appears to break down. A scan of the network graphs for the three periods (see Figures 8.1–8.3) shows that the number of isolates in the insurgent networks, although still substantial during the post-surge period, was much reduced. The clusters of interconnected organizations grew in size and complexity. Indeed, the payoff to building networks may have also changed by this last period: After nearly 10 years of intense, U.S.-led CT/CI efforts, the need for partners to help ensure survival may have driven more organizations together. Conversely, and as noted previously, the Obama administration came to office at the beginning of the post-surge period with a newer, more nuanced vision of CT/CI policy that may have suggested that cautious alliance building by insurgents might be less likely to attract a violent U.S. reaction than in the post-9/11 period.

The remaining effects that assert themselves in only one period all affect the post-surge period specifically. The modeling here suggests just how substantial the change in relational dynamics appears to be after 2007. As noted previously, there were already a number of factors affecting relational development that began in the post-9/11 period and carried forward to the post-surge period, including effects from geographic contiguity, age, state sponsorship, and preference by ethnonationalist organizations to ally with other ethnonationalists. Now, three more effects become significant shapers of the insurgent network globally.

First, in direct contrast with the previous period, in which isolates tended to stay isolates, our modeling suggests that organizations began to build more sprawling networks. As noted previously, there was still a strong tendency toward closed triads, and there was also a pronounced effort to reach out. We found that organizations began to make indirect connections—linkages that require an intermediary or broker. Our visualizations contain more "chains" and linked clusters than in previous periods. Technology may again be a key driver because the continued build-out of the internet made long-distance relationships easier to manage. However, the Obama administration's abandonment of some of the most aggressive CT/CI policies of the Bush administration may also have relieved just enough pressure to make building more extensive networks seem less risky.

Second, the role of ideological homophily grew even more central, as religious organizations also became significantly more likely to ally with one another than with organizations espousing other ideologies. This completes a process of shifting ideological bases for alliance formation over a 15-year period.

Our modeling found that before 9/11, leftists tended to connect with leftists (although the effect is somewhat dubious: $t = 1.789$) and ethnonationalist organizations tended to make fewer alliances (although also with some doubt: $t = 1.840$). The first of these factors—leftists seeking leftists—is probably the last vestige of the dominant ideology of insurgents and terrorists during the Cold War: the various strains of communist ideology that fueled left-wing violence for a couple generations. Substantially older than organizations beholden to other ideologies (Table 8.4), leftist organizations showed a strong affinity for each other, which may reflect both the long-standing nature of the movements and the need to "circle the wagons" as their numbers dwindled and they adjusted to the loss of patronage from the Soviet Union.

By contrast, ethnonationalist movements were beginning to build in number—in fact, the single most common ideology by 1998 was ethnonationalist—but were less long-lived and established than leftists. Religious organizations were even newer and less numerous. In 1998, there were 39 ethnonationalist and 24 religious organizations in our data set. During the pre-9/11 period, only 3 new ethnonationalist and 2 new religious organizations entered the data set. But during the post-9/11 period, 15 ethnonationalist and 9 religious organizations became active.

In the pre-9/11 period, ethnonationalists and religious organizations were simply less likely to seek or be sought for alliance ties. Our modeling suggests, but does not conclusively demonstrate, that the tendency to "go it alone" among ethnonationalists and religious organizations may be related to their relative inexperience at developing relationships (being younger) and the greater risk to established insurgents of seeking or accepting partnerships with less experienced peers. Under these conditions, it may have been safer to go it alone. Thus, although our modeling of *terrorist* organizations found a pronounced tendency for ethnonationalist organizations to ally with one another during the pre-9/11 period (Asal et al. 2016), this tendency did not fully manifest itself among *insurgent* organizations until the post-9/11 period for ethnonationalist organizations and post-surge period for religious organizations. By these later periods, there were more of both ethnonationalist and religious organizations and more with substantial experience. Numbers and experience made alliance formation much more attractive.

The final piece of the puzzle was the emergence of membership size as a relational driver. Large organizations tended to be less involved in relational formation, all else being equal. Like age, there is a logic here of maturity and choice. Large organizations often have deep reservoirs of human and financial

Table 8.4 Average Age of Active Organizations by Period and Ideology

Ideology	Pre-9/11	Post-9/11	Post-Surge
Left	23.32	25.67	32.61
Religious	11.60	10.90	14.74
Ethnonationalist	16.88	16.69	18.98
Other	3.93	3.42	7.26

capital upon which to draw. Given the risks of infiltration, larger organizations may choose to both curtail alliance-seeking and eschew entreaties from interested partners. Although no insurgency is autarkic, larger ones need not take so many relational risks. Larger organizations also tend to be older, and with age come stable sets of supportive, trusted relationships.

8.6. Conclusion

Between 1998 and 2012, the nature of insurgent networks changed dramatically. The global network of insurgent organizations began as a sparse set of ties that formed due to relational closure, proximity, shared competence (via the ability to provide social services and kill people efficiently), and desire to learn how to avoid violent suppression from CT/CI effort. However, by the post-surge period, the network became much larger, more complex, and more mature. The network was formed through shared ideologies, a transition to greater reliance on state sponsorship, and openness to broader alliances that branched out beyond the immediate home base country. The network of 2012 was also a product of the crucible of intense CT/CI pressure during the post-9/11 period, when political terror was adopted widely, forcing some organizations to hunker down as isolates but driving others to find friends that shared ideologies and revenue-generation models such as drug trafficking.

Returning to where we began, there are some continuities in the factors that promote relational formation and help maintain network relationships among insurgent organizations—most prominently, a preference for low-density networks, networks that feature substantial closure (a friend of a friend is also a friend), networks that are local in nature, and connections to peers that are able to accomplish difficult organizational feats such as offering social services. Yet our modeling demonstrates that network formation is very much a product of the political and CT/CI of a particular time. Admittedly, the 15 years that we are studying probably feature more radical change in CT/CI policy, practice, and participation (especially by the United States) since the decolonization wars after World War II and possibly even more so than during that period.

Our findings suggest that the network form of organization is being used by insurgents for precisely the reasons that it has been adopted for domestic provision of public goods and services (Isett et al. 2011): It is flexible and adaptable. The structure and process of network formation morphed along with the environment in ways that hierarchical structures could never achieve. When enacted by skilled network operators, alliance relationships provide substantial adaptive benefits to insurgents that can translate into material and operational benefits, as well as extended conflict and often greater loss of life, as previous chapters have shown. Contrary to what some of the network literature suggests, networks are not an unalloyed good: They are simply one of several organizational technologies that insurgents may select in the pursuit of their mission and goals.

References

Abuza, Zachary. "Tentacles of Terror: Al Qaeda's Southeast Asian Network." *Contemporary Southeast Asia* 24, no. 3 (2002): 427–65.

Abuza, Zachary. "Funding Terrorism in Southeast Asia: The Financial Network of Al Qaeda and Jemaah Islamiya." *Contemporary Southeast Asia* 25, no. 2 (2003): 169–99.

Adamson, Fiona B. "Globalisation, Transnational Political Mobilisation, and Networks of Violence." *Cambridge Review of International Affairs* 18, no. 1 (2005): 31–49.

Arquilla, John. "Networks, Netwar, and Information Age Terrorism." In *Countering the New Terrorism*, edited by Ian O. Lesser, Bruce Hoffman, John Arquilla, David Ronfeldt, and Michele Zanini, 39–84. Santa Monica, CA: RAND, 1999.

Arquilla, John. "Pre-Empting Terror: Take a Networked Approach." *IPA Review* 55, no. 3 (2003): 16–17.

Asal, Victor H., Hyun Hee Park, R. Karl Rethemeyer, and Gary Ackerman. "With Friends Like These . . .: Why Terrorist Organizations Ally." *International Public Management Journal* 19, no. 1 (2016): 1–30.

Asal, Victor H., and R. Karl Rethemeyer. "Pakistan, Kashmir, and Afghanistan: The Role of the Pakistani Inter-Services Intelligence Agency." Unpublished data analysis, Big Allied and Dangerous Database, Version 2.0, 2007.

Asal, Victor, and R. Karl Rethemeyer. "Dilettantes, Ideologues, and the Weak: Terrorists Who Don't Kill." *Conflict Management and Peace Science* 25, no. 3 (2008a): 244–63.

Asal, Victor, and R. Karl Rethemeyer. "The Nature of the Beast: Terrorist Organizational Characteristics and Organizational Lethality." *Journal of Politics* 70, no. 2 (2008b): 437–49.

Asal, Victor H., R. Karl Rethemeyer, Ian Anderson, Allyson Stein, Jeffrey Rizzo, and Matthew Rozea. "The Softest of Targets: A Study on Terrorist Target Selection." *Journal of Applied Security Research* 4, no. 3 (2009): 258–278. doi.org/10.1080/19361610902929990

Barnes, Jamal. "Black Sites, 'Extraordinary Renditions' and the Legitimacy of the Torture Taboo." *International Politics* 53 (2016): 198–219. http://doi.org/10.1057/ip.2015.46

Basile, Mark. "Going to the Source: Why Al Qaeda's Financial Network Is Likely to Withstand the Current War on Terrorist Financing." *Studies in Conflict & Terrorism* 27, no. 3 (2004): 169–85.

Bureau of Investigative Journalism. "Drone Wars: The Full Data." 2020. https://www.thebureauinvestigates.com/stories/2017-01-01/drone-wars-the-full-data.

Burt, Ronald S. *Brokerage and Closure: An Introduction to Social Capital*. New York, NY: Oxford University Press, 2005.

Byman, Daniel. "Why Drones Work: The Case for Washington's Weapon of Choice." *Foreign Affairs* 92, no. 4 (2013): 32–43.

Daly, Max. (2016). "No, Your Drug Use Is Not Funding Terrorism." https://www.vice.com/en_us/article/ppvgw8/is-the-drug-trade-really-bank-rolling-terrorists.

Eizenstat, Stuart E., John Edward Porter, and Jeremy M. Weinstein. "Rebuilding Weak States." *Foreign Affairs* 84, no. 1 (2005): 134–46. http://doi.org/10.2307/20034213.

Farley, Jonathan David. "Breaking Al Qaeda Cells: A Mathematical Analysis of Counterterrorism Operations (A Guide for Risk Assessment and Decision Making)." *Studies in Conflict & Terrorism* 26, no. 6 (2003): 399–411.

Fedotov, Yury. "In Just Two Decades, Technology Has Become a Cornerstone of Criminality." 2017. https://www.huffingtonpost.co.uk/yury-fedotov/in-just-two-decades-techn_b_18330400.html?ncid=engmodushpmg00000004&guccounter=1

Gill, Paul, Jeongyoon Lee, R. Karl Rethemeyer, John Horgan, and Victor H. Asal. "Lethal Connections: The Determinants of Network Connections in the PIRA, 1970–1988." *International Interactions* 40, no. 1 (2014): 52–78. http://doi.org/10.1080/03050629.2013.863190]

Gilmore, Jonathan. "A Kinder, Gentler Counter-Terrorism: Counterinsurgency, Human Security and the War On Terror." *Security Dialogue* 42, no. 1 (2011): 21–37. http://doi.org/10.1177/0967010610393390.

Goderis, Benedikt, and Mila Versteeg. "Human Rights Violations After 9/11 and the Role of Constitutional Constraints." *Journal of Legal Studies* 41, no. 1 (2012): 131–64. http://doi.org/10.1086/663766.

Greenberg, Karen J., and Joshue L. Dratel, eds. *The Torture Papers: The Road to Abu Ghraib*. Cambridge, UK: Cambridge University Press, 2005. http://doi.org/10.1017/CBO9780511511127.

Harding, Luke. "Taliban to Lift Ban on Farmers Growing Opium if US Attacks." *The Guardian*. September 24, 2001. https://www.theguardian.com/world/2001/sep/25/afghanistan.terrorism8.

Hatmaker, Deneen M., and R. Karl Rethemeyer. "Mobile Trust, Enacted Relationships: Social Capital in a State-Level Policy Network." *International Public Management Journal* 11, no. 4 (2008): 426–62. http://doi.org/10.1080/1096749080 2494867.

International Telecommunication Union. "ICT Facts and Figures 2017." 2017. Retrieved October 7, 2018, from https://www.itu.int/en/ITU-D/Statistics/Documents/facts/ICTFactsFigures2017.pdf.

Isett, Kimberley R., Kelly LeRoux, Ines A. Mergel, Pamela A. Mischen, and R. Karl Rethemeyer. "Networks in Public Administration Scholarship: Understanding Where We Are and Where We Need to Go." *Journal of Public Administration Research and Theory* 21, Suppl 1 (2011): i157–73. http://doi.org/10.1093/jopart/muq061.

Jenkins, Brian. "Bush, Obama, and Trump: The Evolution of U.S. Counterterrorist Policy Since 9/11." International Institute for Counter-Terrorism. 2017. https://www.ict.org.il/Article/2079/BUSH-OBAMA-AND-TRUMP#gsc.tab=0

Kenney, Michael. "The Architecture of Drug Trafficking: Network Forms of Organisation in the Colombian Cocaine Trade." *Global Crime* 8, no. 3 (2007): 233–59.

Laqueur, W. *The New Terrorism: Fanaticism and the Arms of Mass Destruction*. New York, NY: Oxford University Press, 1999.

Lee, Jeongyoon, Hyun Hee Park, and R. Karl Rethemeyer. "How Does Policy Funding Context Matter to Networks? Resource Dependence, Advocacy Mobilization, and Network Structures." *Journal of Public Administration Research and Theory* 28, no. 3 (2018): 388–405. http://doi.org/10.1093/jopart/muy016.

Malm, Aili, and Gisela Bichler. "Networks of Collaborating Criminals: Assessing the Structural Vulnerability of Drug Markets." *Journal of Research in Crime and Delinquency* 48, no. 2 (2011): 271–97.

Matthew, Richard, and George Shambaugh. "The Limits of Terrorism: A Network Perspective." *International Studies Review* 7, no. 4 (2005): 617–27.

Mayntz, Renate. *Organizational Forms of Terrorism: Hierarchy, Network, or a Type Sui Generis?* Cologne, Germany: Max Planck Institute for the Study of Societies, 2004.

McAllister, Brad. "Al Qaeda and the Innovative Firm: Demythologizing the Network." *Studies in Conflict & Terrorism* 27, no. 4 (2004): 297–319.

Miller, Abraham H. "The Evolution of Terrorism." *Conflict Quarterly* 5, no. 4 (1985): 5–16.

National Commission on Terrorist Attacks upon the United States. *The 9/11 Commission Report: Final Report of the National Commission on Terrorist Attacks upon the United States*. Washington, DC: National Commission on Terrorist Attacks upon the United States, 2004. https://lccn.loc.gov/2004356401.

Obama, Barack H. "Obama's Remarks on Iraq and Afghanistan." Provided to The New York Times. *The New York Times*. July 15, 2008. https://www.nytimes.com/2008/07/15/us/politics/15text-obama.html.

Omelicheva, Mariya Y., and Lawrence Markowitz. "Does Drug Trafficking Impact Terrorism? Afghan Opioids and Terrorist Violence in Central Asia." *Studies in Conflict & Terrorism* 42, no. 12 (2019): 1021–43. http://doi.org/10.1080/10576 10X.2018.1434039.

Pfeffer, Jeffrey. "A Resource Dependence Perspective on Intercorporate Relations." In *Intercorporate Relations: The Structural Analysis of Business*, edited by M. S. Mizruchi and M. Schwartz, 25–55. New York, NY: Cambridge University Press, 1987.

Pfeffer, Jeffrey, and Gerald R. Salancik. *The External Control of Organizations*. New York, NY: Harper & Row, 1978.

Polo, Sara M. T. "How Terrorism Spreads: Emulation and the Diffusion of Ethnic and Ethnoreligious Terrorism." *Journal of Conflict Resolution* 64, no. 10 (2020): 1916–42.

Popovic, Milos. "Inter-Rebel Alliances in the Shadow of Foreign Sponsors." *International Interactions* 44, no. 4 (2018): 749–76.

Quillen, Chris. "A Historical Analysis of Mass Casualty Bombers." *Studies in Conflict & Terrorism* 25, no. 5 (2002a): 279–92.

Quillen, Chris. "Mass Casualty Bombings Chronology." *Studies in Conflict & Terrorism* 25, no. 5 (2002b): 293–302.

Reid, Edna, Hsinchun Chen, and Jennifer Xu. "Social Network Analysis for Terrorism Research." In *National Security: Handbooks in Information Systems*, edited by Hsinchun Chen, T. S. Raghu, Ram Ramesh, Ajay Vinze, and Daniel Zeng, 243–72. Bingley, UK: Emerald Group, 2007.

Ripley, Ruth M., Tom A. B. Snijders, Zsófia Boda, András Vörös, and Paulina Preciado. "Manual for SIENA Version 4.0 (Version September 18, 2020)." Oxford, UK: University of Oxford, Department of Statistics, Nuffield College, 2020. http://www.stats.ox.ac.uk/snij-ders/siena.

Sageman, Marc. *Understanding Terror Networks*. Philadelphia, PA: University of Pennsylvania Press, 2004.

Schneider, Jane, and Peter Schneider. "The Mafia and al-Qaeda: Violent and Secretive Organizations in Comparative and Historical Perspective." *American Anthropologist* 104, no. 3 (2002): 776–82.

Snijders, Tom A. B. "The Statistical Evaluation of Social Network Dynamics." *Sociological Methodology* 31 (2001): 361–95.

Steinhoff, Patricia G. "Portrait of a Terrorist: An Interview with Kozo Okamoto." *Asian Survey* 16, no. 9 (1976): 830–45.

Steinhoff, Patricia G. "Hijackers, Bombers, and Bank Robbers: Managerial Style in the Japanese Red Army." *Journal of Asian Studies* 48, no. 4 (1989): 724–40.

Stentiford, Barry M. *Success in the Shadows: Operation Enduring Freedom—Philippines and the Global War on Terror, 2002–2015*. Fort Leavenworth, KS: Combat Studies Institute Press, 2018.

Teorell, Jan, Stefan Dahlberg, Sören Holmberg, Bo Rothstein, Anna Khomenko, and Richard Svensson. *The Quality of Government Standard Dataset Codebook*. Gothenburg, Sweden: University of Gothenburg, The Quality of Government Institute, 2016. http://doi.org/10.18157/QoGStdJan16.

Tharoor, Avinash. "ISIS' Brutal Drug War Reflects Global Prohibition Norms." 2016. https://www.talkingdrugs.org/Isis-Brutal-Drug-War-Reflects-Global-Prohibition-Norms.

Turk, Austin T. "Policing International Terrorism: Options." *Police Practice & Research* 3, no. 4 (2002): 279–86.

United Press International. "25 Die at Israeli Airport as 3 Gunmen from Plane Fire on 250 in Terminal." *The New York Times*. May 31, 1972, 2.

Williamson, Oliver E. "The Economics of Organization: The Transaction Cost Approach." *American Journal of Sociology* 87, no. 3 (1981): 548–77. http://doi.org/10.1086/227496.

Zieve, Tamara. "This Week in History: The Lod Airport Massacre." *The Jerusalem Post*. 2012. http://www.jpost.com/features/in-thespotlight/this-week-in-history-the-lod-airport-massacre.

9

Understanding Insurgent Rivalry

9.1. Introduction

In Chapter 3, it was argued that insurgent group interactions are an important part of insurgency dynamics, and the subsequent empirical chapters
(Chapters 4–7) further demonstrated the significance of alliances and rivalry for civilian targeting in particular. Chapter 8 then continued additional
analysis of alliances, showing their changes over time and patterns generally. Given the apparent importance of rivalry, and the data we are presenting
on insurgent organizations, this chapter takes a deeper look at antagonism
among insurgent organizations. The goal is to provide additional descriptive
information on rivalrous relationships and to show the extent to which empirical patterns match up with what the literature suggests we should find.
In addition, we conduct some simple tests to try to explore the explanations
of rivalry. Although the chapter is mostly descriptive, it seeks to provide the
reader with a more nuanced notion of one of our key concepts, interorganizational rivalry, and show initial steps for how future research can proceed.

Unlike Chapter 8, this chapter does not use a network-centric model to analyze changes in rivalry over time. Stochastic actor-oriented models are suitable
for relatively dense networks, but insurgent rivalry does not occur as frequently
as alliances. As a result, this chapter shows rivalrous relationships descriptively,
and it presents the results of group-year analyses to indicate why some groups
are more likely than others to have rivals. This is consistent with most other
global analyses of militant group rivalry (Cunningham, Bakke, and Seymour
2012; Fjelde and Nilsson 2012; Phillips 2019; Warren and Troy 2015).[1]

Descriptively, rivalry patterns change in interesting ways each year. Some
of the most consistently rivalry-prone are Palestinian insurgent organizations. Other persistent clusters are in the Philippines, northeast India, and

---

[1] For an exception, see Gade, Hafez, and Gabbay's (2019) network analysis of insurgent rivalry in
Syria. The Syrian civil war has an unusual number of militant groups involved and an apparently high
rate of rivalry. Another approach, closer to group-year, is the dyad-year unit of analysis (e.g., Conrad,
Greene, Phillips, and Daly 2021).

Insurgent Terrorism. Victor Asal, Brian J. Phillips, and R. Karl Rethemeyer, Oxford University Press. © University
of Maryland National Consortium for the Study of Terrorism and Responses to Terrorism (START) 2022.
DOI: 10.1093/oso/9780197607015.003.0009

Darfur, Sudan. However, members of these clusters change almost every year, and other groups of rivalries occur in other countries throughout the world. The regression results suggest ideology plays a key role in insurgent rivalry. An ethnoreligious separatist ideology is associated with a tripling of the probability of an insurgent group having a rival. This is a larger substantive effect than other factors theorized in the literature to be important, such as involvement in drugs or group strength. Overall, these results are preliminary—an example of what scholars can use Big, Allied, and Dangerous II (BAAD2) Insurgency data for—but they also shed light on the question of insurgent rivalry participation and lay the groundwork for additional work on this important subject.

9.2. Patterns of Rivalry and Dyadic Dynamics

We use NetDraw (Borgatti 2002) to illustrate insurgent networks across time. The BAAD2 network data contain information on three types of rivalry. The first type is factions or splinters of one another (BAAD2 network code no. 3). Per the BAAD2 codebook, factions and splinters form "due to difference in view regarding tactics, ideology, peace talks, goals or other issues. Generally the [factions or splinters] do not maintain positive relationship" (BAAD2 codebook 2018, 21). BAAD2 also records insurgents who are rivals but have no "familial" relationship to one another (BAAD2 network code no. 8). Rival insurgent organizations explicitly compete for the same object or goal, rival insurgencies try to equal or outdo one another, and rival insurgencies consider one another competitors. Rivals seek to dispute one another's preeminence or superiority; Hamas and Fatah are examples. Finally, BAAD2 records when groups engage in insurgent-on-insurgent violence (BAAD2 network code no. 9). Rivalry and alliance are not mutually exclusive in the BAAD2 coding. There are multiple instances in the data in which organizations act both cooperatively and competitively in the same year. This could be due to long-term common interests but tensions over tactics or due to conversion from friendly relations to hostility in the course of a year. Again, Hamas and Fatah are a good example: There are multiple years in which the two organizations act both cooperatively and against the other's interests. For our analysis, the three BAAD2 network codes noted previously (nos. 3, 8, and 9) were each taken as evidence of rivalry. The network data were dichotomized and made symmetric.

For our visualizations, we focus on two characteristics that may help explain the presence or absence of rivalry: ideology and propensity to kill. With respect to ideology, our previous research (Asal et al. 2016) found clear evidence that ideology matters to alliances affiliation. We wanted to probe whether that was also true for rivalry. This previous work found that both religious and ethnonationalist ideologies helped explain alliance formation, but the confluence of both religious and ethnonationalist ideological commitments was particularly powerful. In the current context, because most insurgent organizations that are coded as religiously affiliated are Islamically inspired, we also sought to determine if difference by sect (Sunni vs. Shia) also influences rivalry formation. In addition, we wanted to examine whether support for leftist ideologies, which may be in conflict with some religious tenants, affects rivalry formation. Finally, insurgent organizations may also differ in their commitment to separatism. Indeed, insurgent organizations may differ principally over whether to seek a separate homeland versus some form of autonomy or government participation.

The second dimension that we visualized was the propensity to kill, either through terrorism or through attacks on uniformed military, police, or paramilitary forces. Rivalry could be driven by differences in "effectiveness" as measured by killing. These interests dictated our symbolic language in the visualizations. Each visualization includes only those organizations with one or more rivalry connections, as defined previously. That is, all "isolates" in the rivalry network have been removed. Each organizational symbol (or "node marker") may differ by color, shape, and size. Color indicates ideological commitment to Islam or leftism (see the specific mapping to color in Table 9.1). Organizations that are committed to separatism have a thicker "rim" around the node marker. Finally, larger node markers indicate more killing in a given year, where we anchored the size of the node to the sum of all terrorism [Global Terrorism Database (GTD)] and uniformed forces [Uppsala Conflict Data Program (UCDP)] deaths in a given year. A line between organizations indicates that the two are rivals, per the BAAD2 network coding. Placement in two-dimensional space was initially established using a spring-embedded algorithm, but each visualization was manually revised to improve readability.

In Figures 9.1–9.15, colors indicate ideology. Red is Sunni, maroon is Islamist (no sect), pink is Shia, blue is leftist, and white is none of the above. The size of nodes is based on the lethality of the group, where a larger symbol indicates more people killed by the group in the year, according to

Table 9.1 Key for Figures

Marker	Meaning
Color	
Red	Sunni
Maroon	Islamist (no sect)
Pink	Shia
Blue	Leftist
White	None of the above
Marker size	Larger = more fatalities that year (sources: GTD and UCDP)
Thick border	Separatist ideology
Marker shape	
Diamond	Ethnonationalist
Circle	All others

GTD, Global Terrorism Database; UCDP, Uppsala Conflict Data Program.

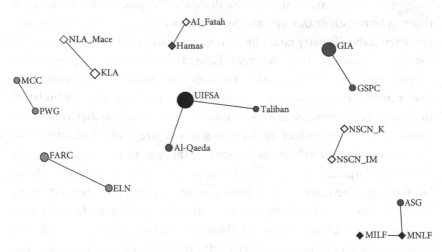

Figure 9.1 Rivalry network, 1998.

the GTD and UCDP data. A black border or rim on the symbol indicates whether the group is separatist or not: A thicker border indicates separatist ideology. Finally, diamond-shaped symbols represent insurgent groups with ethnonationalist ideologies, whereas circles indicate groups that are not ethnonationalist.

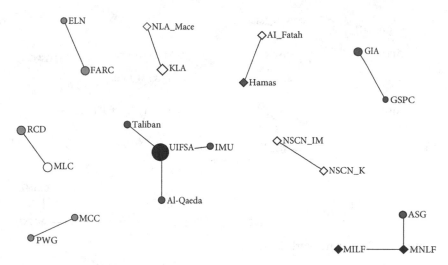

Figure 9.2 Rivalry network, 1999.

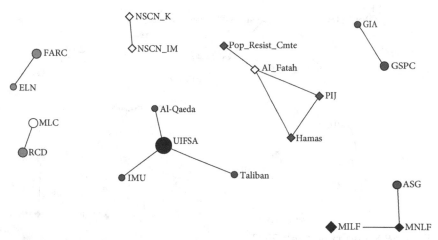

Figure 9.3 Rivalry network, 2000.

Figures 9.1–9.15 show, for each year, which groups had rivals. Some general trends occur. First, a small percentage of groups appear to be rivals in a given year. There are 140 groups in the data, and nearly half (62 or 44%) are in a rivalry at some point. But the rivalries usually do not last for all years. As a result, in the single-year snapshots, we do not see a high rate of rivalry. For example, in 1998, there are 18 groups in rivalries. This increases slightly over the years, but there are always fewer than 30 groups with rivals each year.

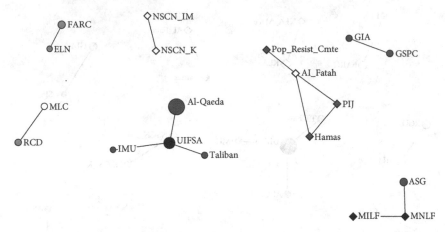

Figure 9.4 Rivalry network, 2001.

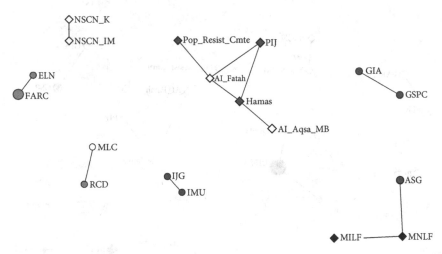

Figure 9.5 Rivalry network, 2002.

The year 2010 is when there were the most insurgent groups in rivalries (29). There were 101 groups in the data that year, so this is a rivalry rate of less than 30%—in the most rivalry-dense year. Overall, rivalries do not occur as frequently as the alliances discussed in the previous chapter.

Second, there is no large network of rivalries. In other words, rivals tend to be pairs of insurgent groups or sometimes as many as five groups, but no more than that. This is quite different from the alliance data in Chapter 8,

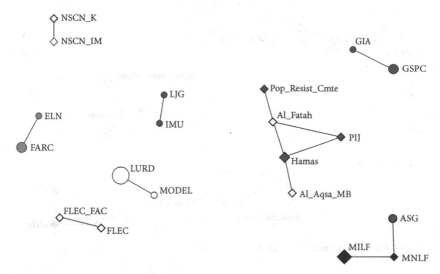

Figure 9.6 Rivalry network, 2003.

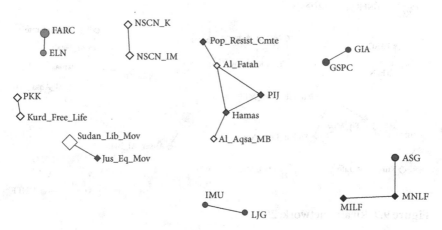

Figure 9.7 Rivalry network, 2004.

which indicated that groups often have multiple allies, and those allies have allies as well. The lack of such network depth among rivals presents methodological limitations: It means we cannot use tools typically applied to relatively dense networks (Corbetta and Grant 2012; Maoz et al. 2007). In practical terms, we cannot study "the enemy of my enemy" if an enemy is unlikely to have additional enemies.

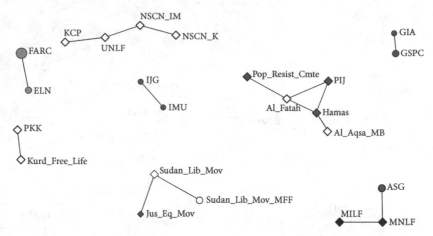

Figure 9.8 Rivalry network, 2005.

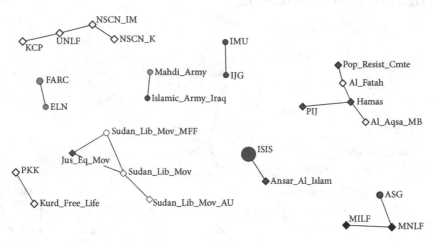

Figure 9.9 Rivalry network, 2006.

Third, of the largest clusters, in most years this is the Israel-based cluster, which has five insurgent groups during 8 years. The core of this cluster seems to be the Fatah versus Hamas rivalry, which is an active rivalry in every year of the data. Other groups join in as rivals of one of these groups, and the Palestinian Jihad is sometimes a rival of both Fatah and Hamas. The only other cluster of five insurgent groups occurs in 2012, with insurgents in Syria. Another relatively large and persistent rival cluster is the Philippines cluster, which involved three groups from 1998 to 2009 and then four groups from

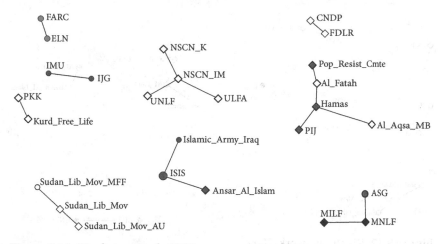

Figure 9.10 Rivalry network, 2007.

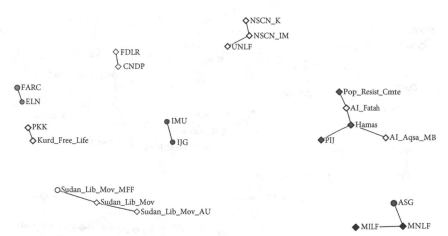

Figure 9.11 Rivalry network, 2008.

2010 to 2012. Interestingly, the Moro National Liberation Front (MNLF) stands out for being the only group in the data with two rivals [Abu Sayyaf Group (ASG) and Moro Islamic Liberation Front (MILF)] for all 15 years of the sample. A set of groups in northeast India are also a persistent cluster, with three or four groups in rivalry in all years of the sample. Finally, several groups in Darfur, Sudan, form a relatively persistent cluster, with two to four groups in rivalry from 2004 to 2012.

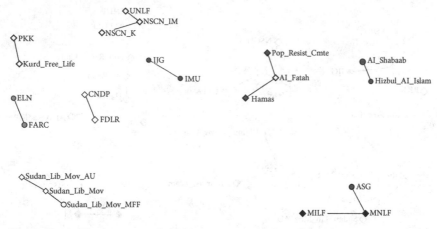

Figure 9.12 Rivalry network, 2009

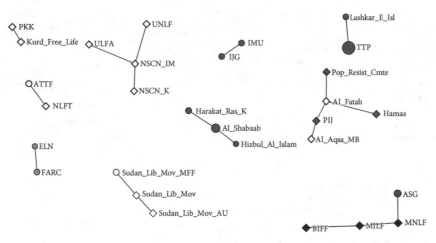

Figure 9.13 Rivalry network, 2010

Fourth, beyond the clusters of three or more groups, there are also a number of dyads of rivals that appear across multiple years. The Revolutionary Armed Forces of Colombia and the National Liberation Army are rivals in all years of the data, joining the few other pairs (Fatah–Hamas, MNLF–MILF, and MNLF–ASG) with that distinction. Other long-running rivalries were those of the Armed Islamic Group and the Salafist Group for Preaching and Combat (1998–2005) in Algeria, several pairs of groups in the Iraq civil conflict, and the Kurdistan Workers Party versus the Kurdistan Free

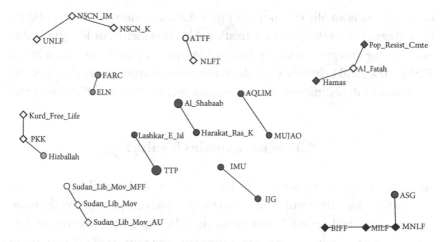

Figure 9.14 Rivalry network, 2011.

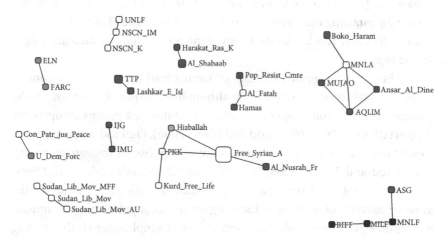

Figure 9.15 Rivalry network, 2012.

Life Party. Other rivalries come and go in the data as short-lived dyads or as single groups joined a rivalry with a group that was already in a rivalry.

Finally, it is noteworthy that in contrast with alliance behavior, most rivalries are local. It seems that generally insurgents are likely to have ties with groups in the same country, but recall that many alliances crossed national boundaries: Afghanistan- or Pakistan-based al-Qaeda aligned with groups in the Philippines and North Africa; Lebanese Hezbollah connected to groups in Iraq, Somalia, and the Palestinian territories; and Nigeria-based

Boko Haram is an ally of groups in other African countries and the Middle East. Regarding rivalry, however, relationships are much more likely to be domestic. One exception is the Syrian civil war rivals, which come from Syria, Turkey, Iraq, and Lebanon. Overall, rivalry is not as common as alliances, but there is still interesting variation across groups, countries/conflicts, and time.

9.3. What Explains Rivalry?

Although many insurgent groups have rivalries, and rivalries seem to be associated with important outcomes (see previous chapters), why do some groups have rivalries while others do not? This question is pertinent because a growing literature seeks to understand the sources of militant group competition. Rivalry is critical to understand because it leads to a great deal of violence (Bloom 2005; Conrad and Greene 2015; Farrell 2020). It is also somewhat puzzling that groups turn on each other instead of their common enemy, the state. This book's data provide an opportunity to shed more light on the topic.

The literature suggests a number of factors lead to rivalry.[2] In a general sense, the literature on outbidding, although not explicitly focused on the question of why groups fight each other, assumes that groups compete for support (Bloom 2004, 2005; Kydd and Walter 2006). Outbidding presupposes rivalry because more extreme tactics are a response to intergroup rivalry (e.g., Kydd and Walter 2006, 76–78). Even Crenshaw's early work (1981, 387) noted that rival terrorist groups attacked civilians to influence the broader movement, instead of focusing on the total public or government. A commonality among all this work is that it emphasizes rivalry among groups claiming to represent the same audience, generally an ethnic group. This is intrafield rivalry (Phillips 2015). According to outbidding research, then, we should expect ethnonationalist groups to be especially rivalrous.

---

[2] This section sets aside the related but distinct literature on insurgent group fragmentation (Doctor 2020; Fjelde and Nilsson 2018; Mosinger 2018; Pearlman and Cunningham 2012; Tamm 2016; Woldemariam 2016). This work is related because fragmentation can lead to rivalry—a group breaks into multiple groups, which might end up fighting each other. These causes of fragmentation, however, seem to be different from the roots of violence between extant insurgent groups. One exception is Cunningham et al. (2012), which mostly examines fragmentation but also examines sources of intergroup violence (see Table 3 in Cunningham et al. 2012). Also note that some fragmentation work examines the fragmentation of entire ethnic groups or movements (Warren and Troy 2015), which could overlap with interinsurgent rivalry. Much of the work on fragmentation, however, examines extant insurgent group fragmentation not necessarily related to rivalry.

Other lines of work that make implicit claims about the sources of rivalry are (1) work on pro-government militias (PGMs) and (2) research on sectarianism. Work on PGMs, if broadly defined, often discusses the natural antagonism between pro-state and anti-state groups. Examples include the loyalist and republican groups in Northern Ireland (Bruce 1992) or right wing and left wing groups in contexts such as Colombia (Romero 2003). Regarding sectarianism, this work often highlights interfield instead of intrafield divisions (Corstange and York 2018). Examples include studies of Sunni versus Shiite rivalry (Abdo 2017; Hashim 2013). Overall, these diverse literatures suggest that rivalry has many causes. One factor that seems to be behind it in several contexts is ethnopolitical motivation. The salience of ethnic identity (as opposed to left–right ideology or general anti-government motivation) seems to suggest a fertile environment for broad movements to consist of multiple militant organizations that fight against each other. This is consistent with the literature on insurgent group fragmentation (e.g., Seymour, Bakke, and Cunningham 2016).

Regarding research specifically on factors leading to intergroup rivalry, some work examines material factors such as financing and territorial control. Berti (2020) qualitatively examines the Syrian civil war and finds that rivalry emerges when the state is absent and resources are scarce. Fjelde and Nilsson (2012) find that insurgent groups are most likely to engage in rivalry when they are involved in drug cultivation, control territory, are relatively strong or weak compared to other insurgents, or operate in a weak state. Phillips (2019) examines the broader category of "terrorist groups," generally including insurgent groups, and finds rivalry associated with involvement in drugs, state sponsorship, ethnic motivation, and operating in a country with a civil conflict. In contrast with Fjelde and Nilsson, Phillips does not find controlling territory, or any (absolute) group size measure, to be associated with rivalry. Stein and Cantin (2021) find that resource attributes such as state sponsorship are helpful for explaining rivalry among rebel groups.

Other work argues that power, and usually relative power, among insurgent groups explains rivalry. Pischedda (2018) argues that relative power combined with co-ethnicity explains rivalry. Specifically, when one group is especially strong and has the opportunity to gain hegemony over other groups, or when one group is dangerously weak, either of these groups might attack their fellow insurgents. Pischedda cautions that power is insufficient for explaining rivalry, however, and argues that co-ethnicity is another necessary part of the explanation. Mendelsohn (2021) suggests that intergroup violence is explained by internal power shifts, spillover from internal conflict in the dominant group, and disagreements over targeting noncombatants. Warren and Troy (2015) argue that relative size differences among ethnic groups in a country encourages ethnic groups to fragment. Gade et al. (2019) find that ideological distance and power asymmetry explain rivalry. Conrad et al. (2021) find power asymmetry is only associated with rivalry among pairs of groups sharing the same ethnic motivation. In contrast with power arguments, Hafez (2020) argues that ideology is key and that groups with ideological distance, yet within the same overarching ideological "family tree," are especially likely to fight.

Empirically, research on rivalry has drawn on a variety of sources of evidence—different from those we use. Pischedda (2018) conducts in-depth qualitative research on insurgent groups in Ethiopia. Hafez (2020) and Mendelsohn (2021) separately examine Algeria's 1990s civil war. Regarding quantitative work, Gade et al. (2019) examine the social media claims of 30 militant groups in Syria. Phillips (2019) examines a global sample of militant groups from 1987 to 2005. Conrad et al. (2021) analyze pairs (dyads) of militant groups in Africa and Asia from 1990–2015.

Fjelde and Nilsson (2012) and Stein and Cantin (2021) have the closest samples to our own, examining insurgent groups between 1989 and 2007, or 1989 and 2018, respectively.[3] Our study offers the advantage of examining more years and insurgent groups than Fjelde and Nilsson, and far more explanatory variables than either study. Overall, although several studies have investigated the sources of militant group rivalry, a variety of arguments have found support with quite different empirical approaches. This relatively new line of research needs more theory and evidence to advance. In the rest of the chapter, we contribute evidence and provide suggestions for additional steps forward.

9.4. Modeling Insurgent Rivalry

Given the previously discussed evidence, we construct a relatively inductive model of insurgent group rivalry to determine if hypotheses from the literature are supported using our data. The outcome of interest is a dichotomous variable

---

[3] See also the data from Powell and Florea (2021), which examines insurgent groups in the Middle East and North Africa, 1993–2018. Powell and Florea do not examine factors associated with rivalry, but one could use their data for that.

Rivalry, as discussed in Chapter 3. The overall data are the same group-year data from 1998–2012 as discussed in the same chapter. We include theoretically relevant independent variables such as *Drugs*, *Territory*, *Group size*, and *GDP per capita (log)*. We also harness our relatively nuanced data to analyze fine-grained data on group ideology, given theoretical interest in the topic. Because of the detailed ideological codings of groups in our data, we are able to examine how specific ideologies might be associated with rivalry more than others. We look at exclusive ideology categories (e.g., leftist, but not ethnic or religious), as well as combined categories such as religious ethnoseparatists. The specific model used is a logistic regression with country random effects to take into consideration the time-series nature of the data. This is consistent with empirical models in previous chapters.

The results shown in Table 9.2 support some arguments in the literature but also contrast with others. Model 1 is a very simple ideology model, including only variables representing four exclusive types of ideologies: leftist, religious, ethnic, and separatist. The omitted category is less common types of ideology, such as right-wing, or groups with multiple ideologies. Model 2 includes combinations of ideologies, such as ethnoreligous groups. This is relevant because some studies have found that combinations such as ethnoreligious are more fruitful for explaining violence compared with single ideologies (Asal and Rethemeyer 2008). Model 3 is the full model, adding the previously mentioned ideology variables with a host of organization-level and country-level variables generally discussed in previous chapters. Model 4 includes even more independent variables, losing parsimony but allowing us to determine if other variables in our data might be related to rivalry. This model is relatively atheoretical—a kitchen sink model—but should be viewed as a potential starting point for future research.

Regarding findings, two exclusive ideology variables, *Religious* and *Separatist*, are positively signed and statistically significant in all models. Groups that are only motivated by religion or separatism are more likely than other types of groups to have a rival. Examples include al-Qaeda in the Islamic Maghreb and the All-Tripura Tiger Force, respectively.[4] Regarding substantive significance, we calculate marginal effects and graph them in Figure 9.16. Religious motivation is associated with a 21 percentage-point increase in the probability of a group having at least one rival in a particular

---

[4] Religious is the most common of the exclusive ideologies because 26 groups are coded for this ideology. Only 4 groups are coded exclusively as separatists.

Table 9.2 Logit Models of Insurgent Group Interorganizational Rivalry

	Model 1	Model 2	Model 3	Model 4
	Simple Ideology Model	Nuanced Ideology Model	Full Model	Additional Variables
Leftist	−0.180	1.462	2.449	2.708
	(1.397)	(1.362)	(1.626)	(1.692)
Religious	3.772**	5.244***	4.562**	4.367**
	(1.408)	(1.335)	(1.508)	(1.625)
Ethnic	−0.511	1.156	1.112	0.997
	(1.325)	(1.326)	(1.600)	(1.632)
Separatist	4.257*	4.833**	6.157**	6.289**
	(2.464)	(2.413)	(2.859)	(2.896)
Ethnoreligious	0.008	−1.005	-0.885	
	(2.638)	(3.112)	(3.160)	
Religious separatist	3.881	4.752*	4.640*	
	(2.473)	(2.769)	(2.805)	
Leftist ethnoseparatist	3.916*	5.177*	5.268*	
	(2.373)	(2.735)	(2.818)	
Religious ethnoseparatist	7.528***	9.014***	8.876***	
	(1.867)	(2.443)	(2.511)	
Alliances		0.190	0.158	
		(0.129)	(0.132)	
Group size		0.073	0.091	
		(0.408)	(0.420)	
Territory		1.577***	1.406**	
		(0.455)	(0.469)	
Group age		−0.026	−0.028	
		(0.035)	(0.037)	
Drugs		1.482**	1.413*	
		(0.717)	(0.721)	
Holding office		−2.699**	−2.434*	
		(1.291)	(1.303)	
Conflict battle deaths		0.000	0.000	
		(0.000)	(0.000)	
Total insurgent groups		1.393**	1.372**	
		(0.455)	(0.458)	

Table 9.2 *Continued*

	Model 1	Model 2	Model 3	Model 4
	Simple Ideology Model	Nuanced Ideology Model	Full Model	Additional Variables
GDP per capita			0.000	0.000
			(0.000)	(0.000)
Democracy			−0.337**	−0.349**
			(0.159)	(0.170)
Carrot				0.157
				(0.992)
Stick				0.178
				(0.353)
Single group leader				0.483
				(1.848)
State sponsorship				0.372
				(0.633)
Charity funding				1.392**
				(0.581)
Constant	−4.879***	−6.305***	−7.812***	−8.049***
	(0.844)	(0.911)	(1.588)	(1.674)
N	1,386	1,386	1,386	1,386

Note: Standard errors in parentheses. Models include group random effects and year fixed effects.
$*p < .10, **p < .05, ***p < .01$.
GDP, gross domestic product.

year. Separatist motivations are associated with a higher increase, 28 percentage points. Neither *Leftist* nor *Ethnic* is ever statistically significant. The
non-finding for ethnic ideology is especially interesting because outbidding
was, at least originally, fundamentally about ethnically motivated groups
or ethnoseparatists (Horowitz 1985; Kaufman 1996; Rabushka and Shepsle
1972). More recent work has shown that ethnic motivation—in a broad, non-
exclusive sense—is related to militant group rivalry as well (Conrad et al.
2021; Fjelde and Nilsson 2012; Phillips 2019).[5] However, our results add nuance to the apparent relationship between ethnicity and insurgent rivalry.

---

[5] Fjelde and Nilsson (2012) use a measure of ethnic mobilization for insurgent groups, which apparently is not an exclusive measure. In other words, it seems it could include ethnoreligious or leftist
ethnic groups as well.

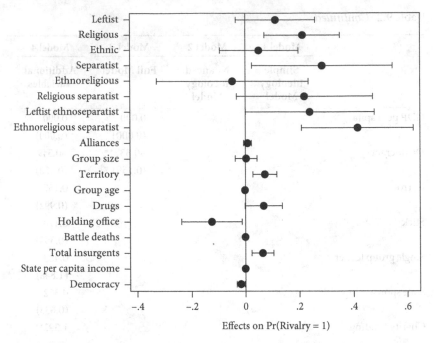

Figure 9.16 Substantive effects from Model 3. Average marginal effects are shown with 95% confidence intervals.

Ethnic motivation alone seems to be insufficient for explaining intergroup confrontations because groups with only ethnic ideologies are not more likely to experience rivalry.

This non-finding dovetails well with findings from the combination ideology variables. These results suggest ethnicity can play an important role—when coalesced with other ideologies. The coefficient on *Ethnoreligious separatist* is positively signed and statistically significant in all three models in which it is included. This suggests an interesting interactive relationship between ethnic and religious motivations. The estimated substantive impact (see Figure 9.16) is that an ethnoreligious separatist ideology is associated with a 42 percentage-point increase in the likelihood of a group having a rival. This is the largest substantive effect in the model. The baseline probability of a group-year being coded for rivalry is approximately 21%, so the difference between a non-ethnoreligious separatist and an ethnoreligious separatist, with other variables held constant, is approximately triple—from 21% to 63%. Two other combination ideology variables, *Ethnoreligious leftist* and *Religious separatist,* are also positively signed and statistically significant in some models, but only at the 10% level. The estimated effects in Model 3 are a 24 or 22 percentage-point increase, respectively, in the likelihood of rivalry. The coefficients on *Ethnoreligious* are never statistically significant, suggesting that this combination of ideologies is not associated with rivalry. This might be surprising given the findings for *Ethnoreligious separatist*; however, very few ethnic groups are not also separatists. As a result, one could say that ethnoreligious motivations are indeed associated with rivalry—as long as these groups are also separatists, which most ethnic insurgent groups are.

Of the other group-level variables in Model 3, the main model, only *Territory, Drugs*, and *Holding office* (negative sign) are associated with rivalry. Interestingly, the substantive effects of these variables are smaller than those of the ideology variables. If a group holds territory, it is 7 percentage points more likely to have a rival than a group not holding territory. The estimated increase is the same if the group is in the drug business. Holding office is associated with a 12 percentage-point decrease in the probability that a group has a rival. These are relatively small effects compared to the 42 percentage-point increase in probably linked to a group having an ethnoreligious separatist ideology.

The statistically significant positive sign on *Territory* is consistent with Fjelde and Nilsson's (2012) results but contrasts with a lack of such a finding by Phillips (2019). It is unclear why territorial control is related to intergroup rivalry. It could be that territory is an indicator of power, and we know that other studies suggest at least relative power matters in important ways for rivalry (e.g., Gade et al. 2019). A related possibility is that territorial control is a visible manifestation of insurgent power, and other groups might want to fight to take territory from those that have it. In addition, groups with territory might seek to expand their control by attacking other territory-occupying insurgents. The result for *Drugs* is consistent with the literature, although note that it is only marginally significant ($p < .10$). It is also remarkable that the effect is substantively small, a 7 percentage-point increase, compared to the ideology variables. The result for *Holding office* is interesting because no one has searched for such a relationship in the past. Groups that have a legal political branch are less likely to be involved in a rivalry with another group. This makes sense because groups operating above the board might avoid confrontations with other militant groups.

Group variables that are not statistically significant are *Allies, Group size*, and *Age*. This is perhaps surprising because all of these variables can be viewed as indicators of opportunity proxies. As groups have more positive interactions with other groups, as the groups grow in size, and the longer they exist, they have more opportunities for negative interactions. This is consistent with explanations of militant group alliances, which often find opportunity indicators such as age to be significant (Asal et al. 2016). Yet these types of opportunities alone do not seem to be sufficient for explaining interorganizational rivalry. *Allies* and *Group size* could also be viewed as indicators of group strength, and a number of studies make power-based arguments for rivalry (Gade et al. 2019; Mendelsohn 2021; Pischedda 2018). These arguments are usually dyadic (relative power among groups) or involve interactions with other factors. It seems that power on its own, however, at least as measured by intergroup alliances and group membership size, does not explain insurgent rivalry. Consistent with this, Gade et al. (2019) find that absolute insurgent group power is not associated with rivalry in Syria, whereas relative power among groups is associated with rivalry.

Regarding state- or conflict-level variables, *Democracy* is negatively signed and statistically significant. Insurgent groups operating in more democratic countries are less likely to have a rival. It is unclear why this might be the case. Some studies find a positive relationship with democracy (Cunningham et al. 2012; Phillips 2019). These studies examine different samples, which could explain part of the difference. Note that some other studies of rivalry have focused on single countries (Gade et al. 2019) or a few cases (Pischedda 2018), so they have not compared country-level characteristics. Exploring the possible relationship between state attributes such as democracy and rivalry is a ripe area for additional research.

Total insurgent groups is statistically significant and positively signed in both models, suggesting that groups in countries with higher numbers of other insurgent groups are more likely to have rivals. Each additional group in an insurgent's country is associated with a 6% increase in the probability that the insurgent has a rival. This makes sense from an opportunity perspective and is consistent with some other work (Cunningham et al. 2012), although it does contrast with a lack of a finding by Fjelde and Nilsson (2012).

State capacity and overall conflict battle deaths are both statistically insignificant. State capacity is apparently important for group–state relations but seems less relevant for intergroup relations. The lack of a significant relationship for conflict battle deaths, which had been important for explaining civilian victimization, is somewhat surprising. One might think that more deadly conflicts also have more interinsurgent fighting. However, probably the majority of battle deaths occur between insurgents and the state, so perhaps it makes sense that such violence is orthogonal to interinsurgent rivalry.

Model 4 includes several variables that were not included in Model 3 because they are not factors clearly suggested by the literature. However, they are mostly variables that have not been included in previous studies, but this is at least in part because extant studies did not have these variables. Because the BAAD2 Insurgency data include these additional variables, we include them to test possible relationships. Of these variables, only *Funded by charity* is positively signed and statistically significant. Note that this refers to insurgents that receive funding from charities, not those that operate them and provide charity services (Flanigan 2006). Not much has been written about why groups that receive charitable donations might also have insurgent enemies. It could well be that, consistent with the outbidding argument, these groups are competing for such donations, and this explains the finding. Insurgent funding is the subject of a number of studies, but much of this work tends to examine state sponsorship, crime, or natural resources (Bapat 2012; San-Akca 2016; Walsh et al. 2018). Other work examines why militant groups provide charity to their community and the consequences of this (Flanigan 2006). More investigation into charity as a funding source for groups, and other potential consequences beyond rivalry, would be fruitful.

Of the other group-level variables in Model 4, *Carrot, Stick, Single leader*, and *State sponsor* are all statistically insignificant. The lack of results for *Carrot* and *Stick*, indicating state concessions or repression to groups (Asal et al. 2019), is notable because one might think that government actions could lead groups to turn on each other. This is apparently not the case. Regarding a single leader, we wondered if a group with less decision-making constraints might be more likely to impulsively attack another insurgent group. This does not seem to be supported by the data, but more research is needed on how leadership types affect insurgent group actions. Finally, some work suggests that state sponsorship can be related to rivalry (Fjelde and Nilsson 2012), but our lack of a finding is consistent with Gade et al.'s (2019) study of Syrian groups.

9.5. Conclusion

This chapter provided an in-depth look at rivalries in the BAAD2 Insurgency data, given that relationships are one of the advantages the data have over other sources. Rivalry also deserves in-depth attention because it is a topic of growing importance as scholars try to understand its causes (Fjelde and Nilsson 2012; Gade et al. 2019; Mendelsohn 2021; Pischedda 2018). The chapter discussed patterns in the data over time, such as the key rivalry clusters in particular years, and showed results of regressions seeking to identify correlates of insurgent rivalry. Regarding general descriptive data on the patterns, rivalry is not a constant occurrence for most groups. Nearly half of the groups had a rival in at least one year, but rivalries sometimes only lasted a single year. Regarding geographic distribution, prominent rivalry clusters appear among Palestinian groups, as well as those in the Philippines, northeast India, and Darfur, Sudan. In general, there is a great deal of variation of time regarding rivalries, as even these prominent clusters gain and lose members each year.

Regression results confirmed some findings from previous studies but also contradicted others. In addition, we tested relationships not yet explored with quantitative data, with interesting results. Some previous work found that holding territory and involvement in the drug trade are associated with rivalry (Fjelde and Nilsson 2012; Phillips 2019). We found the same, but the substantive significance on these variables is quite weak compared to that for ideology-based variables. A new contribution of our study was to examine relatively nuanced measures of group ideology. When examining many types of ideologies, we found that ethnoreligious separatists are especially likely to have rivals. Groups that are motivated by ethnicity alone are not especially likely to have a rival, despite the substantial literature that connects ethnicity to competition and outbidding (e.g., Bloom 2005; Rabushka and Shepsle 1972). Groups with a purely religious motivation—not connected to ethnicity—are also more likely than others to have a rival. Taken together, these results suggest that religion plays a critical role in insurgent rivalry.

We also find that insurgents groups in countries with democratic regimes are less likely to have rivals, which goes against what some other scholars find using a smaller sample (Fjelde and Nilsson 2012). It is unclear why regime type might have such a relationship with insurgent rivalry, but this would be an interesting topic for future research. In addition, we failed to find a relationship between typical group strength measures (group size and group alliances) and rivalry. This is noteworthy because strength and power are perhaps the most common explanations for militant group rivalry (Gade et al. 2019; Mendelsohn 2021; Pischedda 2018). These studies often make relative power arguments, so our lack of a finding highlights the significance of relative versus absolute power.

Going forward, it would be helpful to see more links between the research on insurgent group fragmentation and insurgent group rivalry. Rivalry is an important part of insurgent behavior, as this chapter and book more broadly have sought to show. At the same time, a growing line of work shows important implications of insurgent group fragmentation (Doctor 2020; Fjelde and Nilsson 2018; Mosinger 2018; Pearlman and Cunningham 2012; Tamm 2016; Woldemariam 2016). Some research explores both rivalry and fragmentation (Cunningham et al. 2012), but most work examines them in isolation. Given the growing interest in both topics, it would be fruitful to explore, for example, why fragmentation sometimes leads to rivalry among the resulting groups and sometimes not. In addition, under what conditions does rivalry lead to group fragmentation? Overall, potential relationships with fragmentation add to the intrigue of insurgent rivalry, but regardless, insurgent rivalry is a significant part of insurgency dynamics, and this chapter sought to shed light on the sources of such rivalries. More generally, the topic of militant group rivalry seems to be of growing interest (Berti 2020; Chuang, Ben-Asher, and D'Orsogna 2019; Powell and Florea 2021; Stein and Cantin 2021; Tokdemir et al. 2021). As theorizing develops, and more data are gathered, it will be valuable to better understand this important topic.

References

Abdo, Geneive. *The New Sectarianism: The Arab Uprisings and the Rebirth of the Shi'a-Sunni Divide*. New York, NY: Oxford University Press, 2017.

Asal, Victor H., Hyun Hee Park, R. Karl Rethemeyer, and Gary Ackerman. "With Friends Like These . . .: Why Terrorist Organizations Ally." *International Public Management Journal* 19, no. 1 (2016): 1–30.

Asal, Victor H. and R. Karl Rethemeyer. "The Nature of the Beast: Terrorist: The Organizational and Network Characteristics of Organizational Lethality." *Journal of Politics* 70, no. 2 (2008): 437–49. doi:10.1017/s0022381608080419

Asal, Victor, Brian J. Phillips, R. Karl Rethemeyer, Corina Simonelli, and Joseph K. Young. "Carrots, Sticks, and Insurgent Targeting of Civilians." *Journal of Conflict Resolution* 63, no. 7 (2019): 1710–35.

Bapat, Navin A. "Understanding State Sponsorship of Militant Groups." *British Journal of Political Science* 42, no. 1 (2012): 1–29.

Berti, Bendetta. "From Cooperation to Competition: Localization, Militarization and Rebel Co-Governance Arrangements in Syria." *Studies in Conflict & Terrorism* (2020, June).

Bloom, Mia M. "Palestinian Suicide Bombing: Public Support, Market Share, and Outbidding." *Political Science Quarterly* 119, no. 1 (2004): 61–88.

Bloom, Mia. *Dying to Kill: The Allure of Suicide Terror*. New York, NY: Columbia University Press, 2005.

Borgatti, Stephen P. "NetDraw." Computer program included in UCINET 6 (2002).

Bruce, Steve. *The Red Hand: Protestant Paramilitaries in Northern Ireland*. New York, NY: Oxford University Press, 1992.

Chuang, Yao-Li, Noam Ben-Asher, and Maria R. D'Orsogna. "Local Alliances and Rivalries Shape Near-Repeat Terror Activity of al-Qaeda, ISIS, and Insurgents." *Proceedings of the National Academy of Sciences* 116, no. 42 (2019): 20898–903.

Conrad, Justin, and Kevin T. Greene. "Competition, Differentiation, and the Severity of Terrorist Attacks." *Journal of Politics* 77, no. 2 (2015): 546–61.

Conrad, Justin, Kevin T. Greene, Brian J. Phillips, and Samantha Daly. "Competition from Within: Ethnicity, Power, and Militant Group Rivalry." *Defence and Peace Economics* (2021): Ahead of print.

Corbetta, Renato, and Keith A. Grant. "Intervention in Conflicts from a Network Perspective." *Conflict Management and Peace Science* 29, no. 3 (2012): 314–40.

Corstange, Daniel, and Erin A. York. "Sectarian Framing in the Syrian Civil War." *American Journal of Political Science* 62, no. 2 (2018): 441–55.

Crenshaw, Martha. "The Causes of Terrorism." *Comparative Politics* 13, no. 4 (1981): 379–99.

Cunningham, Kathleen Gallagher, Kristin M. Bakke, and Lee J. M. Seymour. "Shirts Today, Skins Tomorrow: Dual Contests and the Effects of Fragmentation in Self-Determination Disputes." *Journal of Conflict Resolution* 56, no. 1 (2012): 67–93.

Doctor, Austin C. "A Motion of No Confidence: Leadership and Rebel Fragmentation." *Journal of Global Security Studies* 5, no. 4 (2020): 598–616.

Farrell, Megan. "The Logic of Transnational Outbidding: Pledging Allegiance and the Escalation of Violence." *Journal of Peace Research* 57, no. 3 (2020): 437–51.

Fjelde, Hanne, and Desirée Nilsson. "Rebels Against Rebels: Explaining Violence Between Rebel Groups." *Journal of Conflict Resolution* 56, no. 4 (2012): 604–28.

Fjelde, Hanne, and Desiree Nilsson. "The Rise of Rebel Contenders: Barriers to Entry and Fragmentation in Civil Wars." *Journal of Peace Research* 55, no. 5 (2018): 551–65.

Flanigan, Shawn Teresa. "Charity as Resistance: Connections Between Charity, Contentious Politics, and Terror." *Studies in Conflict & Terrorism* 29, no. 7 (2006): 641–55.

Gade, Emily Kalah, Mohammed M. Hafez, and Michael Gabbay. "Fratricide in Rebel Movements: A Network Analysis of Syrian Militant Infighting." *Journal of Peace Research* 56, no. 3 (2019): 321–35.

Hafez, Mohammed M. "Fratricidal Rebels: Ideological Extremity and Warring Factionalism in Civil Wars." *Terrorism and Political Violence* 32, no. 3 (2020): 604–29.

Hashim, Ahmed S. *Iraq's Sunni Insurgency*. New York, NY: Routledge, 2013.

Horowitz, Donald L. *Ethnic Groups in Conflict*. Berkeley, CA: University of California Press, 1985.

Kaufman, Stuart J. "Spiraling to Ethnic War: Elites, Masses, and Moscow in Moldova's Civil War." *International Security* 21, no. 2 (1996): 108–38.

Kydd, Andrew H., and Barbara F. Walter. "The Strategies of Terrorism." *International Security* 31, no. 1 (2006): 49–80.

Maoz, Zeev, Leslie G. Terris, Ranan D. Kuperman, and Ilan Talmud. "What Is the Enemy of My Enemy? Causes and Consequences of Imbalanced International Relations, 1816–2001." *Journal of Politics* 69, no. 1 (2007): 100–115.

Mendelsohn, Barak. "The Battle for Algeria: Explaining Fratricide Among Armed Nonstate Actors." *Studies in Conflict & Terrorism* 44, no. 9 (2021): 776–98.

Mosinger, Eric S. "Brothers or Others in Arms? Civilian Constituencies and Rebel Fragmentation in Civil War." *Journal of Peace Research* 55, no. 1 (2018): 62–77.

Pearlman, Wendy, and Kathleen Gallagher Cunningham. "Nonstate Actors, Fragmentation, and Conflict Processes." *Journal of Conflict Resolution* 56, no. 1 (2012): 3–15.

Phillips, Brian J. "Enemies with Benefits? Violent Rivalry and Terrorist Group Longevity." *Journal of Peace Research* 52, no. 1 (2015): 62–75.

Phillips, Brian J. "Terrorist Group Rivalries and Alliances: Testing Competing Explanations." *Studies in Conflict & Terrorism* 42, no. 11 (2019): 997–1019.

Pischedda, Costantino. "Wars Within Wars: Why Windows of Opportunity and Vulnerability Cause Inter-Rebel Fighting in Internal Conflicts." *International Security* 43, no. 1 (2018): 138–76.

Powell, Stephen R., and Adrian Florea. "Introducing the Armed Nonstate Actor Rivalry Dataset (ANARD)." *Civil Wars* (2021). doi:10.1080/13698249.2021.1883334.

Rabushka, Alvin, and Kenneth Shepsle. *Politics in Plural Societies*. Columbus, OH: Merrill, 1972.

Romero, Mauricio. *Paramilitares y Autodefensas*. Bogotá, Columbia: IEPRI-Planeta, 2003.

San-Akca, Belgin. *States in Disguise: Causes of State Support for Rebel Groups*. New York, NY: Oxford University Press, 2016.

Seymour, Lee J. M., Kristin M. Bakke, and Kathleen Gallagher Cunningham. "E Pluribus Unum, Ex Uno Plures: Competition, Violence, and Fragmentation in Ethnopolitical Movements." *Journal of Peace Research* 53, no. 1 (2016): 3–18.

Stein, Arthur, and Marc-Olivier Cantin. "Crowding Out the Field: External Support to Insurgents and the Intensity of Inter-Rebel Fighting in Civil Wars." *International Interactions* 47, no. 4 (2021): 662–91.

Tamm, Henning. "Rebel Leaders, Internal Rivals, and External Resources: How State Sponsors Affect Insurgent Cohesion." *International Studies Quarterly* 60, no. 4 (2016): 599–610.

Tokdemir, Efe, Evgeny Sedashov, Sema Hande Ogutcu-Fu, Carlos E. Moreno Leon, Jeremy Berkowitz, and Seden Akcinaroglu. "Rebel Rivalry and the Strategic Nature of Rebel Group Ideology and Demands." *Journal of Conflict Resolution* 65, no. 4 (2021): 729–58.

Walsh, James Igoe, Justin M. Conrad, Beth Elise Whitaker, and Katelin M. Hudak. "Funding Rebellion: The Rebel Contraband Dataset." *Journal of Peace Research* 55, no. 5 (2018): 699–707.

Warren, T. Camber, and Kevin K. Troy. "Explaining Violent Intra-Ethnic Conflict: Group Fragmentation in the Shadow of State Power." *Journal of Conflict Resolution* 59, no. 3 (2015): 484–509.

Woldemariam, Michael. "Battlefield Outcomes and Rebel Cohesion: Lessons from the Eritrean Independence War." *Terrorism and Political Violence* 28, no. 1 (2016): 135–56.

10

Conclusion

10.1. Introduction

Why examine the killing of civilians by insurgent organizations? Scholars and policymakers seek to understand brutality against civilians, given civilians' unique status compared to combatants and members of the security forces more broadly. A great deal of research, including previous work by the authors of this book, has examined factors explaining terrorism and the dynamics behind terrorist groups—often defined as groups that primarily use terrorism, and usually thought of as too weak to directly attack military troops. However, it is not only terrorist groups that attack civilians. Insurgent organizations—groups fighting government security forces—frequently turn their weapons on noncombatants. Despite the reputation of terrorism as something that happens "randomly" in otherwise peaceful situations, some analysts find that the majority of terrorism happens as a part of civil conflict processes (Findley and Young 2012a).

When we think of terrorism, perhaps the first image in our minds is of the attacks of September 11, 2001 (9/11), when al-Qaeda killed nearly 3,000 people. Al-Qaeda is normally described as a terrorist organization, but it and its branches have been insurgent organizations—fighting government soldiers in combat—for many years. For example, in 2001, in addition to killing so many civilians, al-Qaeda also fought the U.S. military in Afghanistan, which resulted in 1,585 "battle deaths" (Pettersson, Högbladh, and Öberg 2019). Combat between al-Qaeda and the U.S. coalition in Afghanistan killed another 743 people the following year. Indeed, for most of the years covered by the data described in this book, al-Qaeda was involved in combat with security forces in Afghanistan, killing at least 100 people each year. However, as discussed in this book, not all organizations that engage in killing civilians also kill security personnel, and not all organizations that kill security personnel kill civilians. Why then, do some insurgent organizations target civilians, whereas others do not?

Insurgent Terrorism. Victor Asal, Brian J. Phillips, and R. Karl Rethemeyer, Oxford University Press. © University of Maryland National Consortium for the Study of Terrorism and Responses to Terrorism (START) 2022.
DOI: 10.1093/oso/9780197607015.003.0010

Many observers describe civilian targeting as a weapon of the weak, and important research suggests a lack of group capacity may indeed help explain terrorism (Hultman 2007; Polo and Gleditsch 2016). Other valuable scholarship suggests government attributes such as democracy encourage civilian targeting (Chenoweth 2010), including by insurgent organizations (Hultman 2012; Stanton 2013). In this book, we presented an alternate framework, which we describe as insurgent embeddedness—emphasizing an insurgent group's relationships with the state, other insurgents, and the public. This framework, we have argued, helps us understand civilian targeting. In this concluding chapter, we summarize our findings, consider policy recommendations, and discuss new directions for future research related to insurgent organizations and violence against civilians.

10.2. Summarizing the Argument and Findings

In Chapter 1, it was noted that in the past 20 years, there has been a large and growing literature on attacks against civilians by terrorist organizations and, increasingly, by insurgent organizations (Eck and Hultman 2007; Hultman 2012; Keels and Kinney 2019; Stanton 2013). Some of the literature has specifically examined insurgent organizations but with a focus largely on the impact of the features of the states in which they are operating (although not only that) on their behavior (Abrahms and Potter 2015; Hultman 2007; Humphreys and Weinstein 2006; Polo and Gleditsch 2016; Wood 2010; Wood, Kathman, and Gent 2012). Much of the existing literature does not examine the relationship between the specific features of insurgent organizations, their other behaviors, or their connections to other insurgent organizations. As we have noted, this primary focus on relations with states misses out on the key impact that the actual nature and behavior of the insurgent organization might have on its decision to specifically target civilians.

Chapter 2 delved into the different theoretical perspectives that we planned to use to analyze the question of when insurgent organizations kill civilians, eventually presenting our embeddedness theory of civilian victimization. Our theoretical basis starts with the overlapping frameworks of social network analysis (Granovetter 1985; Wasserman and Faust 1994) and embeddedness theory (Polanyi 1944). We agree with Granovetter's (1985, 487) emphasis that actors are "embedded in concrete, ongoing systems of social relations." We argued that insurgent group relations with the state, other insurgents, and the public should be crucial for encouraging or discouraging terrorism. Building on our recent work (Asal et al. 2019), we argued that government concessions to insurgent groups should decrease the likelihood of civilian targeting, whereas repression should increase subsequent civilian targeting. We also argued that alliances and rivalries should both lead to a greater chance of terrorism, through a number of causal mechanisms. Regarding relationships with the public, we focused on three indicators: insurgent social service provision, ethnic ideology, and involvement in crime. These factors indicate specific kinds of interactions with the broader public, all of which are likely to encourage civilian abuse.

Chapter 3 introduced the Big, Allied, and Dangerous II (BAAD2) Insurgency data set. The data set offers several key contributions. First, its focus on insurgent organizations is critical for understanding the environment in which most civilian victimization occurs—civil conflict. It also allows us to study relatively comparable groups—all of those that are or have been involved in such conflict. A second major contribution of the data set is its yearly variation. Some other data collection efforts only have one value for group size, for example, covering all years that a group exists. Our more fine-grained data set allows researchers to examine changes over time to try to determine their causes and consequences. Finally, the BAAD2 Insurgency data set includes many variables unavailable in any other source, such as government concessions or repression toward specific groups, interorganizational relationships, leadership types, and nuanced ideology measures.

Chapters 4–7 tested the theoretical model, exploring how the seven indicators of our model—government concessions and coercion, interorganizational alliances and rivalry, social service provision, ethnic motivations, and crime—might be related to civilian targeting. Each chapter used a different type of civilian targeting as the outcome of interest. The results are summarized in Table 10.1. Chapter 4 examined violence against civilians, finding government concessions associated with a lower likelihood of civilian victimization and all other factors associated with a greater chance of such victimization. These results are robust to many changes in model specification. Chapter 5 explored why some groups mostly attack members of the general public—individuals who are not, for example, government employees or religion or business representatives. Government coercion, rivalry, and ethnic motivation were connected to this type of violence, consistent with our expectations.

Table 10.1 Comparing Relationships Across Different Dependent Variables: Chapters 4–7

	Attacks on Civilians (Chapter 4)	Attacks on the General Public (Chapter 5)	Attacks on Schools (Chapter 6)	Attacks on Journalists (Chapter 7)
Hypothesized Variables				
Carrot	Reduced likelihood	N.S.	No observations	N.S.
Stick	Increased likelihood	Increased likelihood	Increased likelihood	N.S.
Alliances	Increased likelihood	N.S.	Increased likelihood	Increased likelihood
Rivalry	Increased likelihood	Increased likelihood	N.S.	N.S.
Social services	Increased likelihood	N.S.	Increased likelihood	N.S.
Ethnic motivation	Increased likelihood	Increased likelihood	N.S.	N.S.
Crime	Increased likelihood	N.S.	N.S.	N.S.
Selected Other variables				
Territory	N.S.	N.S.	N.S.	Increased likelihood
Democracy	N.S.	Increased likelihood	N.S.	Increased likelihood

Note: Table shows the relationship between the independent variables and different dependent variables. Relationships refer to main models, Model 3 in each chapter, and at least 90% statistical significance.

N.S., not significant.

Chapter 6 examined attacks on educational institutions. This type of target has rarely been studied, and it overlaps in interesting ways with the notion of civilian victimization. It is an extreme type of target because children are some of the most vulnerable people. We find that government coercion, insurgent alliances, and social service provision are associated with these kinds of attacks. Finally, Chapter 7 applied the model to another interesting—and, unfortunately, understudied—type of civilian victimization: violence against journalists. Of the variables in our theoretical framework, only insurgent alliances were associated with this type of attack. We speculated that this could be because militant groups learn about new types of violence, and how to conduct it, via their allies. Interestingly, we also found territorial control and democracy related to attacks on the news media. However, few factors were related to this type of violence, perhaps because it is so rare.

Looking across all the results in Table 10.1, of the seven variables representing aspects of the embeddedness argument, none are associated with all four types of violence. This suggests different logics explain each type of targeting—there is no one factor that can explain so many distinct types of terrorism. Regarding factors associated with multiple outcomes, government coercion ("stick") is linked with subsequent attacks on civilians generally, the general public, and schools. This adds to the growing body of work demonstrating that government repression can backfire, discussed more later. Intergroup alliances are associated with subsequent attacks on civilians generally, schools, and journalists. These are the only two variables in the model associated with so many different types of violence. Government coercion and interorganizational alliances, then, are especially important for understanding insurgent violence against civilians of various types. It is noteworthy that interorganizational rivalries are associated with both civilian targeting and groups focusing on the general public. This provides support for the outbidding argument—the empirical validity of which is frequently debated (Farrell 2020; Findley and Young 2012b).

Chapters 8 and 9 took a different empirical turn, drilling down on the topics of insurgent alliances and rivalry. Given that the theoretical focus of the book is on relationships, and our data offer unique opportunities to shed light on these relationships in particular, we explore these two types of interorganizational relationships in detail. We used stochastic actor-oriented models to identify factors that make an organization more likely to form an alliance and keep that alliance with other insurgents. The analyses look at three distinct periods: pre-9/11, post-9/11, and what we call the post-surge period, starting in 2008 after the increase in U.S. troops in Iraq. Interestingly, alliance connections in general for insurgent organizations are fairly rare. Although some organizations have a large number of connections, many have none. In general, older organizations and organizations that have state sponsors are less likely to have allies at all. Perhaps not surprisingly, groups created closed relationships – among insurgent groups, a friend of a friend is a friend. Organizational features also clearly have an impact on whether a group is likely to have connections. Social service provision makes an organization more likely to have connections, as does territorial control. Not all features and behaviors, however, have the same impact across all time periods. Being involved in terrorist attacks against civilians made an organization more likely to have alliances before the post-surge period but not after. Following the attacks on 9/11, several factors had an impact both immediately after the attacks and in the final post-surge period. Ethnonationalist organizations have been more likely to ally with each other after 9/11. In general, Chapter 8 demonstrated that the situation that an organization finds itself in and the behavior in which it is engaged are likely to have an impact on whether it has alliances and how many.

Chapter 9 studied the factors that seem to explain why some groups have insurgent rivals. Unlike for alliances, this chapter did not use stochastic modeling because rivalries are relatively rare compared to alliances, so such methods are not possible. We built a simple model of rivalry, emphasizing ideology. We found that religious and separatist groups are likely to have rivals, but the type of ideology most likely to lead to rivalry is the combination of religious ethnoseparatist. Other factors include territorial control and the drug trade. Holding office is associated with a lower likelihood of rivalry, whereas democracy, perhaps surprisingly, is linked to a lower chance of rivalry.

10.3. Policy Implications

Although our intention of writing this book was to contribute to our academic understanding of when insurgent organizations target civilians, there are important policy implications to our work as well. Many analysts consider terrorism and insurgency as distinct phenomena, but our book has instead emphasized the overlaps. Of the 140 groups in our data, 38 do not seem to use terrorism, but 87 insurgent groups use terrorism for at least 1 year during our study period (1998-2012). A minority of the groups (15 – just over 10%) use terrorism every single year. Leaving aside our analysis of the how and why, this finding is important in its own right because it highlights that insurgent organizations should not be classified homogeneously as actors that use violence in similar ways. Doing so obfuscates important differences not only in terms of analysis but also regarding how governments should be interacting with these groups—as well as examining the factors that can push organizations toward killing civilians. Our quantitative analysis of why some insurgent organizations are more likely to attack civilians dives deeper into important policy choices that states make and should, we believe, be taken into consideration as states determine how they will deal with insurgents.

Regarding specific policies states could use to address insurgent organizations, most studies examine specific tactics such as leadership targeting (Jordan 2014; Tominaga 2018) or the effects of repression in general (Danneman and Ritter 2014; Piazza 2017; Walsh and Piazza 2010). Our nuanced analyses, which examine additional state strategies, have important implications for how states might confront insurgent organizations more effectively. A twin set of findings from our book—those of state coercion and concessions—suggest government actions have the potential to profoundly affect the use of terrorism by insurgents. One of the most robust relationships in our analyses was that of government coercion and subsequent insurgent terrorism. Government coercion was often followed by attacks on civilians generally, schools, and the general public. This might give government pause about crackdowns. This is especially the case when considering the related finding that government *concessions* are usually associated with a *lower* likelihood of civilian targeting the following year. Regarding attacks on schools in particular, there were simply no cases of groups receiving concessions in one year and attacking a school the following year. Negotiations with insurgent groups are politically sensitive and can backfire. But given the growing body of work suggesting that repression often leads to counterproductive results, such as increased violence (e.g., Belgioioso 2018), perhaps governments should give more thought to alternate conflict-resolution mechanisms.

In addition to specific government approaches we analyzed, some of our other findings have important implications for how policymakers might reduce civilian victimization. Two types of insurgent group interactions—alliances and rivalry—are both consistently related to subsequent civilian victimization. Alliances can help groups aggregate capability and learn new destructive tactics, whereas rivalry leads to terrorism through a number of pathways. Governments and other stakeholders could watch out for, and possibly try to dissuade, insurgent alliances and rivalry as a way to mitigate civilian victimization. Regarding rivalry, this would mark a substantial change in approach, as some governments have encouraged competition among insurgent groups.[1]

10.4. Limitations and Future Research

Future research could build on our work in a number of ways. Two key limitations of our work in the preceding chapters could be addressed. The first is the limited time period covered by this book. We examined only 15 years, mostly during one international context (the post-Cold War era, mostly during the "War on Terror"). The second key limitation is that we study insurgent organizations. To deal with the first limitation, an obvious step would be to expand the time line both forward and backward to determine if differences across global environments have a meaningful impact on the use of violence. Regarding the second limitation, it could be valuable to gather data on multiple types of violence perpetrated by non-state actors. Before this book project, our research into groups and violence began with the coding of terrorist organizations (Asal and Rethemeyer 2008). A key next step is to code both terrorist and insurgent organizations and use the expanded data to look at when violent non-state actors across the definitional divide choose who they target for violence.

There is another important limitation that is not addressed in this book or by the suggestions discussed previously. We started the book with a discussion of the 1989 terrorist attack on a bus in Israel (Zieve 2012). One issue we did not focus on is why someone would kill for political reasons in the first place (Gurr 1970; Lichbach 1998). Why did Abd al-Hadi Ghanim get on that bus to kill civilians? Why did his organization, Palestinian Islamic Jihad, choose violence over nonviolence? Our primary focus in this book has been on why organizations choose one form of violence over another. At this very moment, millions of people around the world are being oppressed, some are being tortured, and others are being killed by their governments. Most of these people do not pick up weapons and start using violence to resist.

---

[1] For example, the commander of U.S. forces in Afghanistan stated, "The Taliban are fighting ISIS, and we encourage that because ISIS needs to be destroyed" (Seldin 2018).

Although there is a great deal of work on the choice to resist either through violence or through protest (Gamson 1990; Gurr and Moore 1997; Lichbach 1998; Pearlman 2011; Saxton 2005; Saxton and Benson 2006; Scarritt and McMillan 1995), very little of this work focuses on organizations and organizational features and behaviors over time the way we do in this book. One exception to this is the use of the Minorities at Risk Organizational Behavior Project data set (2009) to analyze the choice of violence, protest, or neither by ethnopolitical organizations in the Middle East and North Africa (Asal et al. 2013). There is also a growing and important line of work using data on nonviolent campaigns. However, this work also focuses on country attributes (Cunningham et al. 2017) or the attributes of campaigns—more loosely organized movements, as opposed to groups (Belgioioso 2018). It could be fruitful to examine a broad collection of organizations, such as political parties and clubs, to determine which organizations eventually took up arms.

In addition to limitations that may be addressed, another way to build on the work of this book is to use some noteworthy findings as starting points. For example, we repeatedly found social service provision to be associated with civilian victimization and attacks against educational institutions in particular. A growing line of research examines social service provision (Heger and Jung 2017; Huang and Sullivan 2021) and the related idea of governance by militant groups (Arjona, Kasfir, and Mampilly 2015; Stewart 2018), but more work is needed in this area. What are the specific causal mechanisms linking social service provision to violence, including extreme violence? There are many kinds of social services, such as health care and education. Which are most likely to be associated with violence against civilians, and why?

Regarding dependent variables, we examined why some groups focus their attacks on what we called the "general public," people who are not government employees or symbols such as religious leaders. We do not know of any studies that have examined this type of terrorism. Although "randomness" has been described as a fundamental part of terrorism (Schinkel 2009; Walzer 1977), a great deal of civilian targeting is on leaders and symbolic targets, not the general public. More work should examine this particular type of civilian targeting—excluding government civilians in particular—because it seems to have quite different explanations than civilian targeting that also includes government employees. Distinctions among target types can help us better understand the logic(s) behind civilian targeting.

We also found interesting results regarding violence against schools and attacks on journalists. There have been few studies on these targets (Gohdes and Carey 2017; Holland Rios 2017; Petkova et al. 2017), despite their normative importance and theoretical puzzles, such as why groups would carry out such heinous attacks. Do our results hold up in samples of terrorist groups or in other time periods? What other factors explain such extreme targeting? What are the consequences of this kind of violence? There are other types of civilian targets that could be analyzed to identify potential links with embeddedness indicators such as intergroup alliances and rivalries. Some types of targets that could be studied as outcomes include election-related facilities (Birch, Daxecker, and Höglund 2020; Fjelde and Höglund 2021), financial institutions (Liargovas and Repousis 2010), and tourists (Enders, Sandler, and Parise 1992; Walters, Wallin, and Hartley 2019). Overall, civilian targeting is critical to understand. We hope this book not only uncovers interesting findings about victimization of civilians but also provides paths for additional related research.

References

Abrahms, Max, and Philip B. K. Potter. "Explaining Terrorism: Leadership Deficits and Militant Group Tactics." *International Organization* 69, no. 2 (2015): 311–42.
Arjona, Ana, Nelson Kasfir, and Zachariah Mampilly, eds. *Rebel Governance in Civil War.* Cambridge University Press, 2015.
Asal, Victor, Richard Legault, Ora Szekely, and Jonathan Wilkenfeld. "Gender Ideologies and Forms of Contentious Mobilization in the Middle East." *Journal of Peace Research* 50, no. 3 (2013): 305–18.
Asal, Victor, Brian J. Phillips, R. Karl Rethemeyer, Corina Simonelli, and Joseph K. Young. "Carrots, Sticks, and Insurgent Targeting of Civilians." *Journal of Conflict Resolution* 63, no. 7 (2019): 1710–35.
Asal, Victor, and R. Karl Rethemeyer. "Dilettantes, Ideologues, and the Weak: Terrorists Who Don't Kill." *Conflict Management and Peace Science* 25, no. 3 (2008): 244–63.
Belgioioso, Margherita. "Going Underground: Resort to Terrorism in Mass Mobilization Dissident Campaigns." *Journal of Peace Research* 55, no. 5 (2018): 641–55.
Birch, Sarah, Ursula Daxecker, and Kristine Höglund. "Electoral Violence: An Introduction." *Journal of Peace Research* 57, no. 1 (2020): 3–14. https://doi.org/ 10.1177%2F0022343319889657.
Chenoweth, Erica. "Democratic Competition and Terrorist Activity." *Journal of Politics* 72, no. 1 (2010): 16–30.
Cunningham, David E., Kristian Skrede Gleditsch, Belén González, Dragana Vidović, and Peter B. White. "Words and Deeds: From Incompatibilities to Outcomes in Anti-Government Disputes." *Journal of Peace Research* 54, no. 4 (2017): 468–83.

Danneman, Nathan, and Emily Hencken Ritter. "Contagious Rebellion and Preemptive Repression." *Journal of Conflict Resolution* 58, no. 2 (2014): 254–79.

Eck, Kristine, and Lisa Hultman. "One-Sided Violence Against Civilians in War: Insights from New Fatality Data." *Journal of Peace Research* 44, no. 2 (2007): 233–46.

Enders, Walter, Todd Sandler, and Gerald F. Parise. "An Econometric Analysis of the Impact of Terrorism on Tourism." *Kyklos* 45, no. 4 (1992): 531–54.

Farrell, Megan. "The Logic of Transnational Outbidding: Pledging Allegiance and the Escalation of Violence." *Journal of Peace Research* 57, no. 3 (2020): 437–51.

Findley, Michael G., and Joseph K. Young. "Terrorism and Civil War: A Spatial and Temporal Approach to a Conceptual Problem." *Perspectives on Politics* 10, no. 2 (2012a): 285–305.

Findley, Michael G., and Joseph K. Young. "More Combatant Groups, More Terror?: Empirical Tests of an Outbidding Logic." *Terrorism and Political Violence* 24, no. 5 (2012b): 706–21.

Fjelde, Hanne, and Kristine Höglund. "Introducing the Deadly Electoral Conflict Dataset (DECO)." *Journal of Conflict Resolution* (2021, June).

Gamson, William A. *The Strategy of Social Protest*. Belmont, CA: Wadsworth, 1990.

Gohdes, Anita R., and Sabine C. Carey. "Canaries in a Coal-Mine? What the Killings of Journalists Tell Us About Future Repression." *Journal of Peace Research* 54, no. 2 (2017): 157–74.

Granovetter, Mark. "Economic Action and Social Structure: The Problem of Embeddedness." *American Journal of Sociology* 91, no. 3 (1985): 481–510.

Gurr, Ted R. *Why Men Rebel*. Princeton, NJ: Princeton University Press, 1970.

Gurr, Ted R., and Will Moore. "Ethnopolitical Rebellion: A Cross-Sectional Analysis of the 1980s with Risk Assessments for the 1990s." *American Journal of Political Science* 41, no. 4 (1997): 1079–1103.

Heger, Lindsay L., and Danielle F. Jung. "Negotiating with Rebels: The Effect of Rebel Service Provision on Conflict Negotiations." *Journal of Conflict Resolution* 61, no. 6 (2017): 1203–29.

Holland, Bradley E., and Viridiana Rios. "Informally Governing Information: How Criminal Rivalry Leads to Violence Against the Press in Mexico." *Journal of Conflict Resolution* 61, no. 5 (2017): 1095–1119.

Huang, Reyko, and Patricia L. Sullivan. "Arms for Education? External Support and Rebel Social Services." *Journal of Peace Research* 58, no. 4 (2021): 794–808.

Hultman, Lisa. "Battle Losses and Rebel Violence: Raising the Costs for Fighting." *Terrorism and Political Violence* 19, no. 2 (2007): 205–22.

Hultman, Lisa. "Attacks on Civilians in Civil War: Targeting the Achilles Heel of Democratic Governments." *International Interactions* 38, no. 2 (2012): 164–81.

Humphreys, Macartan, and Jeremy M. Weinstein. "Handling and Manhandling Civilians in Civil War." *American Political Science Review* 100, no. 3 (2006): 429–47.

Jordan, Jenna. "Attacking the Leader, Missing the Mark: Why Terrorist Groups Survive Decapitation Strikes." *International Security* 38, no. 4 (2014): 7–38.

Keels, Eric, and Justin Kinney. "'Any Press Is Good Press?' Rebel Political Wings, Media Freedom, and Terrorism in Civil Wars." *International Interactions* 45, no. 1 (2019): 144–69.

Liargovas, Panagiotis, and Spyridon Repousis. "The Impact of Terrorism on Greek Banks' Stocks: An Event Study." *International Research Journal of Finance and Economics* 51 (2010): 88–96.

Lichbach, Mark. "Contending Theories of Contentious Politics and the Structure–Action Problem of Social Order." *Annual Review of Political Science* 1, no. 1 (1998): 401–24.

Minorities at Risk Organizational Behavior Project. *Minorities at Risk Organizational Behavior (MAROB) Code Book.* College Park, MD: Center for International Development and Conflict Management, 2009.

Pearlman, Wendy. *Violence, Nonviolence, and the Palestinian National Movement.* Cambridge, UK: Cambridge University Press, 2011.

Petkova, Elsiveta P., Stephanie Martinez, Jeffrey Schlegelmilch, and Irwin Redlener. "Schools and Terrorism: Global Trends, Impacts, and Lessons for Resilience." *Studies in Conflict & Terrorism* 40, no. 8 (2017): 701–11.

Pettersson, Therese, Stina Högbladh, and Magnus Öberg. "Organized Violence, 1989–2018 and Peace Agreements." *Journal of Peace Research* 56, no. 4 (2019): 589–603.

Piazza, James A. "Repression and Terrorism: A Cross-National Empirical Analysis of Types of Repression and Domestic Terrorism." *Terrorism and Political Violence* 29, no. 1 (2017): 102–18.

Polanyi, Karl. (1944). *The Great Transformation: Economic and Political Origins of Our Time.* New York, NY: Rinehart.

Polo, Sara M. T., and Kristian Skrede Gleditsch. "Twisting Arms and Sending Messages: Terrorist Tactics in Civil War." *Journal of Peace Research* 53, no. 6 (2016): 815–29.

Saxton, Gregory D. "Repression, Grievances, Mobilization, and Rebellion: A New Test of Gurr's Model of Ethnopolitical Rebellion." *International Interactions* 31, no. 1 (2005): 87–116.

Saxton, Gregory D., and Michelle A. Benson. "Structure, Politics, and Action: An Integrated Model of Nationalist Protest and Rebellion." *Nationalism and Ethnic Politics* 12, no. 2 (2006): 137–75.

Scarritt, James R., and Susan McMillan. "Protest and Rebellion in Africa: Explaining Conflicts Between Ethnic Minorities and the State in the 1980s. *Comparative Political Studies* 28, no. 3 (1995): 323–49.

Schinkel, Willem. "On the Concept of Terrorism." *Contemporary Political Theory* 8, no. 2 (2009): 176–98.

Seldin, Jeff. "IS in Afghanistan Just Won't Go Away, US Officials Say." Voice of America. August 17, 2018. https://www.voanews.com/south-central-asia/afghanistan-just-wont-go-away-us-officials-say.

Stanton, Jessica A. "Terrorism in the Context of Civil War." *Journal of Politics* 75, no. 4 (2013): 1009–22.

Stewart, Megan. "Megan Stewart." *International Organization* 72, no. 1 (2018): 205–26.

Tominaga, Yasutaka. "Killing Two Birds with One Stone? Examining the Diffusion Effect of Militant Leadership Decapitation." *International Studies Quarterly* 62, no. 1 (2018): 54–68.

Walsh, James I., and James A. Piazza. "Why Respecting Physical Integrity Rights Reduces Terrorism." *Comparative Political Studies* 43, no. 5 (2010): 551–77.

Walters, Gabrielle, Ann Wallin, and Nicole Hartley. "The Threat of Terrorism and Tourist Choice Behavior." *Journal of Travel Research* 58, no. 3 (2019): 370–82.

Walzer, Michael. *Just and Unjust Wars: A Moral Argument with Historical Illustrations.* New York, NY: Basic Books, 1977.

Wasserman, Stanley, and Katherine Faust. *Social Network Analysis: Methods and Applications.* Vol. 8. Cambridge, UK: Cambridge University Press, 1994.

Wood, Reed M. "Rebel Capability and Strategic Violence Against Civilians." *Journal of Peace Research* 47, no. 5 (2010): 601–14.

Wood, Reed M., Jacob D. Kathman, and Stephen E. Gent. "Armed Intervention and Civilian Victimization in Intrastate Conflicts." *Journal of Peace Research* 49, no. 5 (2012): 647–60.

Zieve, T. "This Week in History: Terror Attack on Bus 405." *The Jerusalem Post.* July 1, 2012. Retrieved July 19, 2020, from https://www.jpost.com/Features/In-Thespotlight/This-Week-In-History-Terror-attack-on-Bus-405.

Index

For the benefit of digital users, indexed terms that span two pages (e.g., 52–53) may, on occasion, appear on only one of those pages.

Note: Tables and figures are indicated by *t* and *f* following the page number.

Abu Sayyaf, 109, 180, 196–97
  criminal behavior, 36
  drug trafficking, 177
  journalists, attacks on, 135, 140
  terrorism, frequency of, 52–53
Algeria, insurgent rivalries in, 198–99
alliances, among insurgent groups
  consequences of, 11
  drug trafficking, 177–79
  ethnonationalist movements, 182
  group size, 182–83
  home-country alliances, 164–65, 170
  ideology, 170–71, 181–82, 182t
  isolates and, 180–81
  leftist motivation, 181–82
  policy implications, 221–22
  political terror, 179–80
  as research variable, 54–55, 66–67, 67t
  state sponsorship, 175
  and targeting of general public, 92–93
  and targeting of journalists, 137–38
  and targeting of schools, 113–14
  violence, 176–77
  See also insurgent alliances, longitudinal
    modeling of
All-Tripura Tiger Force (ATTF), 96, 203–6
analysis, of attacks against schools,
    119–26, 120f
  discussion, 123–26
  results, 120–23, 121t, 123f
analysis, of civilian targeting, 13, 14–15
Armed Islamic Group (GIA), Algeria,
    52–53, 90, 198–99
Azawad National Liberation Movement
  rivalries, 55
  targeting of journalists, 134–35

Baloch Liberation Army
  targeting of general public, 90
  targeting of journalists, 134–35
Beslan school siege, Russia, 108
Biberman, Yelena, 124
Big, Allied, and Dangerous II
    (BAAD2) Insurgency data, 11–12, 45,
    60, 217
  antecedents and first steps, 45–49
  description of, 12–13
  identifying organizations to
    include, 47–49
  number of groups and years in data
    sample, 50f
  overview of data, 49–60
  regional variations in, 49, 50f
  variables, 51–60, 51t, 52f,
    53f, 58f
Bloom, Mia, 114, 116–17, 124
Boko Haram, 29, 199–200
  targeting of general public, 87, 90
  targeting of schools, 108, 118
  2011 violence in Nigeria, 4

charities, funding by, 209
Chechen separatists, and Beslan school
    siege, 108
civil conflict
  definition of, 7
  insurgent embeddedness and,
    25f, 25–26
civilian targeting
  analysis of, 13, 14–15
  of general public, 13
  of journalists, 14
  of schools, 13–14

civilian targeting, theoretical perspectives
on, 12, 21–22, 36–37
embeddedness theory, 23–25
insurgent embeddedness, 23–36
insurgent embeddedness and civil
conflict, 25f, 25–26
insurgents and the public, 32–36
insurgents and the targeted state, 27–29
relationships between insurgent
groups, 29–32
civilian victimization, state-based, 9
coercion, insurgent groups and, 27–29
and attacks on general public, 92
and attacks on journalists, 137
and attacks on schools, 112–13
policy implications, 221
Committee to Protect Journalists, 132
concessions to insurgent groups, 27–29
and attacks on general public, 91
and attacks on journalists, 137
and attacks on schools, 112
policy implications, 221
criminal involvement
and insurgents' relationship with the
public, 36
as research variable, 56–57, 67–69, 68t
and targeting of general public, 96
and targeting of journalists, 140–41
and targeting of schools, 117–18, 127–28

Darfur, insurgent rivalries in, 189–90,
196–97, 209–10
data sources, description of, 45, 60
antecedents and first steps, 45–49
identifying organizations to
include, 47–49
number of groups and years in data
sample, 50f
overview of data, 49–60
regional variations in, 49, 50f
variables, 51–60, 51t, 52f, 53f, 58f
democracy
as control variable, 72–73
and rivalry among insurgent
groups, 208
and targeting of journalists, 147–48
density, of insurgency networks, 163–64
diffusion, and use of terror among
insurgent groups, 166–68, 167t

drone strikes, use by governments, 176–77
drug trafficking
and alliances among insurgent
groups, 177–79
and rivalries among insurgent
groups, 207
dyadic dynamics, and patterns of rivalry,
190–200
rivalry network, 1998, 192f
and rivalry network, 1999, 193f
and rivalry network, 2000, 193f
and rivalry network, 2001, 194f
and rivalry network, 2002, 194f
and rivalry network, 2003, 195f
and rivalry network, 2004, 195f
and rivalry network, 2005, 196f
and rivalry network, 2006, 196f
and rivalry network, 2007, 197f
and rivalry network, 2008, 197f
and rivalry network, 2009, 198f
and rivalry network, 2010, 198f
and rivalry network, 2011, 199f
and rivalry network, 2012, 199f

education, as target type, 108–10, 109t
See also schools, targeting of
embeddedness theory, 23–25
ethnic motivations, 34–35
and causes of rivalry, 203–7
as research variable, 56, 67–69, 68t
and targeting of general public, 95–96
and targeting of journalists, 140
and targeting of schools, 116–17
ethnonationalist movements, and alliances
among insurgent groups, 182
examples of insurgent terrorism, 3–5

Fatah, rivalry with Hamas, 196–97
Fjelde, Hanne, 203–9, 205n.5
Free Aceh Movement, targeting of
journalists, 134
Free Syrian Army, targeting of general
public, 90

general public, and insurgents, 32–36
general public, targeting of, 13, 85–87,
102–3, 217
broader audience for, 89–90
data and analysis on, 96–102, 98f

and definition of terrorism, 85–86
ethnic motivations and, 95–96
future research, 223
government employees and
  property, 88
and insurgent relationships, 92–94
randomness of, 87–90
reasons for, 90–96, 91t
research results, 98–102, 99t, 101f
role of the state, 91–92
social service provision and, 94–95
government employees and property,
  targeting of, 88
gross domestic product (GDP), as control
  variable, 73
group size
  and alliances among insurgent
    groups, 182–83
  as research variable, 58f, 58–59

Hamas, 47–48, 159–60
  alliance embeddedness, 54–55
  motivaions, 56
  rivalry with Fatah, 30, 94, 190, 196–97
  rivals, number of, 55
  social services, provision of, 55–56
  suicide terrorism, 116–17
  targeting of general public, 95
  targeting of journalists, 138–39
  targeting of schools, 109
hypotheses, testing primary, 65, 81, 217
  alternate independent variables, 79t
  count dependent variables, 77t
  dependent variables and alternate
    dependent variables, 74t
  estimation of regression models,
    69–81, 70t
  preliminary analysis and contingency
    tables, 65–69, 66t, 67t, 68t
  results, 69–81
  substantive significance of variables, 72f
  support for hypotheses across models
    and variables, 81t

ideology
  and alliances among insurgent groups,
    170–71, 172f, 173f, 174f, 181–82, 182t
  and patterns of rivalry, 191
  as research variable, 56, 57–58

and rivalry network, 1998, 192f
and rivalry network, 1999, 193f
and rivalry network, 2000, 193f
and rivalry network, 2001, 194f
and rivalry network, 2002, 194f
and rivalry network, 2003, 195f
and rivalry network, 2004, 195f
and rivalry network, 2005, 196f
and rivalry network, 2006, 196f
and rivalry network, 2007, 197f
and rivalry network, 2008, 197f
and rivalry network, 2009, 198f
and rivalry network, 2010, 198f
and rivalry network, 2011, 199f
and rivalry network, 2012, 199f
India, insurgent rivalries in, 196–97
insurgency network structure,
    determinants of, 163–84
  density, 163–64
  factors consistent across all study
    periods, 163–65
  factors consistent across two study
    periods, 166–75
  factors that appear, disappear, and
    reappear, 175–79
  factors unique to particular
    periods, 179–83
  home-country alliances, 164–65, 170
  ideology, 172f, 173f, 174f, 181–82, 182t
  social service provision, 165
  terrorism, and alliances among militant
    groups, 166–69
  transitive triads, 164
insurgent alliances, longitudinal modeling
    of, 155–57, 183–84, 219–20
  data collection, 159–60
  data modeling, 160–63
  determinants of network
    structure, 163–83
  factors consistent across all study
    periods, 163–65
  factors consistent across two study
    periods, 166–75
  factors that appear, disappear, and
    reappear, 175–79
  factors unique to particular
    periods, 179–83
  microstructures, 160
  and microstructures, 158

insurgent alliances, longitudinal
     modeling of (*cont.*)
  Stochastic Actor-Oriented Models
     (SOAMs), 157–58
  terrorism, and alliances among
     insurgent groups, 166–69
  time periods covered, 161–63
insurgent embeddedness
  argument and findings, 216–20
  context of, 215–16
  definition of, 5–6
  and educational-institution targeting,
     111–19, 119t
  limitations and future research, 222–24
  and news media targeting, 136–41
  policy implications, 220–22
  relational context, 7–8
  relationships across dependent
     variables, 218t
insurgent embeddedness, and civilian
     victimization, 23–37
  embeddedness theory, 23–25
  insurgent embeddedness and civil
     conflict, 25f, 25–26
  insurgents and the public, 32–36
  insurgents and the targeted state, 27–29
  relationships between insurgent
     groups, 29–32
insurgent organization
  age and forming alliances, 171–75
  definition of, 6
  vs. militant group, 6
  vs. terrorist organization, 6–7
insurgent rivalries, 15, 31–32
  and attacks on journalists, 138–39
  and attacks on schools, 114–15
  consequences of, 11
  explanations for, 200n.2
  localized nature of, 199–200
  policy implications, 221–22
  as research variable, 54–55, 66–67, 67t
  and targeting of general public, 93–94
insurgents
  and other insurgent groups, 29–32
  and the public, 32–36
insurgent terrorism
  examples of, 3–5
  key concepts, 6–8

Islamic State (ISIS), 162
  attacks on general public, 87
  attacks on religious institutions, 110
  combat against government soldiers, 5
  drug trafficking, 178
isolates, and alliances among insurgent
     groups, 180–81
Israel, 47–48, 89n.4, 155
  insurgent rivalries, 196–97
  Palestinian Islamic Jihad 1989 bus
     attack, 3

Japanese Red Army (JRA), 155–56
journalists, targeting of, 14, 131–33,
     148–49, 219
  and alliances among insurgent
     groups, 137–38
  and coercion toward insurgent
     groups, 137
  and concessions to insurgent
     groups, 137
  and criminal involvement, 140–41
  discussion of research, 145–48
  empirics, 141–48, 142f
  and ethnic motivations, 140
  future research, 224
  insurgent embeddedness and news
     media targeting, 136–41
  news media as target type, 133–35,
     133n.3, 133n.4
  research results, 142–46, 143t, 144f
  and rivalry among insurgent
     groups, 138–39

Kurdistan Workers Party (PKK),
     116–17
  attacks on school facilities, 116
  criminal enterprises, 101, 103
  rivalry with Kurdistan Free Life
     Party, 198–99
  targeting of general public, 95

leadership type, as research variable, 59
leftist motivation
  and alliances among insurgent
     groups, 181–82
  and causes of rivalry, 203–7
  as research variable, 57–58

longitudinal modeling of insurgent
    alliances, 155–57, 183–84, 219–20
data collection, 159–60, 159t
data modeling, 160–63
determinants of network
    structure, 163–83
factors consistent across all study
    periods, 163–65
factors consistent across two study
    periods, 166–75
factors that appear, disappear, and
    reappear, 175–79
factors unique to particular
    periods, 179–83
microstructures, 160
and microstructures, 158
Stochastic Actor-Oriented Models
    (SOAMS), 157–58
terrorism, and alliances among
    insurgent groups, 166–69
time periods covered, 161–63
Lord's Resistance Army (LRA), 85, 90

media, targeting of
and insurgent embeddedness, 136–41
media as target type, 133–35,
    133n.3, 133n.4
Memorial Institute for the Prevention of
    Terrorism (MIPT), 45–46
microstructures, and insurgent alliances,
    158, 160
militant group vs. insurgent group, 6
Moro Islamic Liberation Front (MILF),
    8, 196–97
Moro National Liberation Front
    (MNLF), 196–97

National Democratic Front of
    Bodoland, 95–96
National Liberation Army (ELN),
    Colombia, 11, 29, 54–55, 115
drug trafficking, 177
rivalries, 198–99
targeting of journalists, 131, 134–35
news media, targeting of
and insurgent embeddedness, 136–41
news media as target type, 133–35,
    133n.3, 133n.4

Nigeria, Boko Haram 2011 violence, 4
Nilsson, Desirée, 203–9, 205n.5
Northern Ireland, rivalries among
    insurgent groups in, 115

Obama, Barack, 176–77
office holding, and rivalry among
    insurgent groups, 207

Pakistani Inter-Services Intelligence
    agency (ISI), 175
Palestine, Popular Front for the Liberation
    of, 155–56
Palestinian Islamic Jihad, 1989 bus
    attack, 3
People's United Liberation Front (PULF),
    India, 125
Philippines, insurgent rivalries in, 196–97
Pischedda, Costantino, 201–2
policy implications, of insurgent
    embeddedness, 220–22
political office holding, and rivalry among
    insurgent groups, 207
political terror (state-sanctioned terror)
    and alliances among insurgent
    groups, 179–80
as research variable, 59–60
Popular Front for the Liberation of
    Palestine (PFLP), 155–56
population size, as research
    variable, 59–60
power, and causes of rivalry, 201–2
primary hypotheses, testing, 65, 81
alternate independent variables, 79t
count dependent variables, 77t
dependent variables and alternate
    dependent variables, 74t
estimation of regression models,
    69–81, 70t
preliminary analysis and contingency
    tables, 65–69, 66t, 67t, 68t
results, 69–81
substantive significance of variables, 72f
support for hypotheses across models
    and variables, 81t
pro-government militias, and causes of
    rivalry, 201
public, and insurgents, 32–36

public, targeting of, 13, 85–87, 102–3, 217
    broader audience for, 89–90
    data and analysis on, 96–102, 98f
    and definition of terrorism, 85–86
    ethnic motivations and, 95–96
    future research, 223
    government employees and property, 88
    and insurgent relationships, 92–94
    randomness of, 87–90
    reasons for, 90–96, 91t
    research results, 98–102, 99t, 101f
    role of the state, 91–92
    social service provision and, 94–95

al-Qaeda, 56–57
    drug trafficking, 177, 178
    history of insurgency, 215
al-Qaeda in the Arabian Peninsula, 95

randomness, of violence against general
    public, 87–90
Real Irish Republican Army (RIRA),
    criminal involvement of, 125
regime type, as research variable, 59–60
regional variation, in data sources, 49, 50f
relational context
    and insurgent embeddedness, 7–8
    and targeting of general public, 92–94
religion, and causes of rivalry, 203–7
religious institutions, targeting of, 110
religious motivation, as research variable,
    57–58, 72–73
research, state of relevant, 8–11
Revolutionary Armed Forces of
    Columbia (FARC)
    rivalry with National Liberation Army
        (ELN), 115, 198–99
    targeting of general public, 95
    targeting of journalists, 134–35
    targeting of schools, 107, 109
Revolutionary United Front, in Sierra
    Leone, 21–22
rivalry, among insurgent groups, 15, 31–32
    and attacks on journalists, 138–39
    and attacks on schools, 114–15
    consequences of, 11
    explanations for, 200n.2
    localized nature of, 199–200

policy implications, 221–22
as research variable, 54–55, 66–67, 67t
and targeting of general public, 93–94
rivalry, understanding insurgent, 189–90,
    209–11, 220
    dyadic dynamics, 190–200
    explanations for, 200–2
    modeling insurgent rivalry, 202–9,
        204t, 206f
    patterns of rivalry, 190–200
    rivalry network, 1998, 192f
    and rivalry network, 1999, 193f
    and rivalry network, 2000, 193f
    and rivalry network, 2001, 194f
    and rivalry network, 2002, 194f
    and rivalry network, 2003, 195f
    and rivalry network, 2004, 195f
    and rivalry network, 2005, 196f
    and rivalry network, 2006, 196f
    and rivalry network, 2007, 197f
    and rivalry network, 2008, 197f
    and rivalry network, 2009, 198f
    and rivalry network, 2010, 198f
    and rivalry network, 2011, 199f
    and rivalry network, 2012, 199f
    RSIENA, 158

Salafist Group for Preaching and
    Combat, 198–99
Schinkel, Willem, 88–89
schools, targeting of, 13–14, 107–8, 126–
    28, 219
    and alliances among insurgent
        groups, 113–14
    and concessions to insurgent
        groups, 112
    criminal involvement and, 117–18,
        127–28
    discussion of research, 123–26
    education as target type, 108–10, 109t
    empirical analysis of, 119–26, 120f,
        121t, 123f
    and ethnic motivations, 116–17
    future research, 224
    and government coercion, 112–13
    hypotheses, 119t
    insurgent embeddedness and,
        111–19, 119t

insurgents and the targeted
state, 112–13
research results, 120–23, 121*t*, 123*f*
and rivalries among insurgent
groups, 114–15
and social service provision, 115–16,
127
separatism, and causes of rivalry, 203–7
al-Shabaab, targeting of journalists,
131, 134
Sierra Leone, Revolutionary United Front
in, 21–22
Snijders, Tom A. B., 158, 161
social service provision, 33–34
as factor in network alliances, 165
future research on, 223
as research variable, 55–56,
67–69, 68*t*
and targeting of general public, 94–95
and targeting of journalists, 139–40
and targeting of schools, 115–16, 127
state-based civilian victimization, 9
state sponsorship, and alliances among
insurgent groups, 175
Stochastic Actor-Oriented Models
(SOAMs), 157–58
and convergence, 160
and statistical significance, 163
and time heterogeneity, 160–63
success, terrorism as marker of, 169
Sudan, insurgent rivalries in, 196–97
Sunni and Shi´ite groups, rivalries
among, 115
Syria, insurgent rivalries, 196–97

tactical diffusion, 31, 31n.2
and use of terror among insurgent
groups, 166–68, 167*t*
Taliban
drug trafficking, 177–79
ethnic motivations and targeting of
general public, 95–96
targeting of journalists, 131,
131n.2

targeted state
insurgents and the, 27–29
schools and, 112–13
Tehrik-i-Taliban Pakistan, 107
territorial control
and causes of rivalry, 201, 207
and civil conflict, 10
as research variable, 57
terrorism
and alliances among insurgent groups,
166–69, 167*t*
definition of, 7, 85–86
hard *vs.* soft targets, 109–10, 109*t*
as marker of success, 169
as research variable, 53*f*, 53, 67–69
terrorist organization, *vs.* insurgent
organization, 6–7
theoretical perspectives, on civilian
targeting, 12, 21–22, 36–37, 216–17
embeddedness theory, 23–25
insurgent embeddedness, 23–36
insurgent embeddedness and civil
conflict, 25*f*, 25–26
insurgents and the public, 32–36
insurgents and the targeted state, 27–29
relationships between insurgent
groups, 29–32
transitive triads, insurgency networks
and, 164

Uganda, Lord's Resistance Army in, 85
Uppsala Conflict Data Program (UCDP),
7, 12–13, 22, 47, 159, 191

violence
and alliances among insurgent
groups, 176–77
choosing over nonviolence, 223
key incentives for using, 8–9

Walzer, Michael, 88–89
Williamson, Oliver, 171–75

Zahid, Farhan, 124

Printed in the USA/Agawam, MA
January 21, 2022

788092.019